Finding Your Research Voice

Itai Cohen • Melanie Dreyer-Lude

Finding Your Research Voice

Story Telling and Theatre Skills for Bringing Your Presentation to Life

Itai Cohen
Department of Physics
Cornell University
Ithaca, New York, USA

Melanie Dreyer-Lude
Department of Drama
University of Alberta
Edmonton, Alberta, Canada

ISBN 978-3-030-31519-1   ISBN 978-3-030-31520-7   (eBook)
https://doi.org/10.1007/978-3-030-31520-7

This Springer imprint is published by the registered company Springer Nature Switzerland AG.
The registered company address is: Gewerbestrasse 11, 6330 Cham, Switzerland

Acknowledgments

We would like to extend special thanks to Susi Varvayanis, the Executive Director of the BEST program at Cornell University, for her unwavering support of our work. Susi offered sage advice in each of the workshops we provided for her program and helped shape the content of this book. We thank Sara Leanne Eddleman for her keen eye in helping to edit this manuscript. We also want to thank the postdocs and graduate students who were participants. They were key players in helping us learn how best to help them deliver exceptional research presentations.

Contents

About the Authors

Itai Cohen is a Professor of Physics at Cornell University, where he works on materials in motion. His research topics have ranged from studying the behavior of shear thickening fluids like corn-starch, to the flight of insects, to microscale robots, and the behaviors of crowds. Professor Cohen has given over 250 invited public, conference, and departmental speaking engagements. He has chaired the American Physical Society Forum on Outreach and Engaging the Public and organized numerous professional development workshops on science communication.

Melanie Dreyer-Lude is an artist-scholar specializing in international, intercultural, and interdisciplinary research. She has directed, produced, and taught theater in Chile, Argentina, Germany, Turkey, Uganda, Greece, and Canada. Fluent in German, Dreyer-Lude translates and directs contemporary German plays, which have been produced in the USA and Canada and published in international magazines and anthologies. She currently serves as Chair of the Department of Drama at the University of Alberta and lives and works in Edmonton, Alberta, Canada.

Cohen and Dreyer-Lude are the organizers and instructors of the popular Finding Your Scientific Voice workshop, which has been run at various venues including Cornell University, SUNY Upstate, the Howard Hughes Medical Institute, and the American Physical Society March Meeting.

1

Introduction

When is the last time you heard a compelling, interesting, or memorable research talk? Conferences are notorious for providing ample opportunity to see boring presentations of what could be important research. If you are reading this book, chances are that you (or your students) need to learn how to tell a better story. Whether you have been giving lackluster presentations at professional conferences, you are on the job market and need to present your portfolio to a search committee, or you have found yourself tongue-tied in the middle of an ideal networking opportunity, learning to tell a compelling research story can have a significant impact on your career. Humans enjoy telling and listening to great stories. These stories help us make sense of the world around us.

Good stories demand attention. Bad stories put an audience to sleep. Good stories provide emotional and intellectual satisfaction. Bad stories frustrate an audience and feel like a waste of time. The components of a great story signal when we should pay close attention and where we will find important moments. By learning to tell a compelling research story, you can trigger an emotional connection to your audience, which will help them remember the important information you have just presented. It is easy to tell a research story badly. It takes time and effort to learn to tell one well. Once you have mastered the basic principles of good storytelling, you will experience the satisfaction of performing successfully in front of an audience.

You can find ample reference materials on presenting talks as stories. We invite you to read them. Afterwards you may find there is still a gap between understanding what these books are telling you and incorporating the concepts into *your own* work. This book fills that gap, presenting a range of key

© Springer Nature Switzerland AG 2019
I. Cohen, M. Dreyer-Lude, *Finding Your Research Voice*,
https://doi.org/10.1007/978-3-030-31520-7_1

research presentation techniques, followed by field-tested exercises that will help you improve *your* talk. To demonstrate, let's look at an example.

The Elevator Pitch

A good elevator pitch provides a concise description of your research and why it matters. A compelling elevator pitch will consist of a few clear sentences that include:

1. What are you researching?
2. Why is this problem important?
3. What have others done, and why was that approach not sufficient?
4. What you are doing differently to solve the problem?
5. If you are successful, how will your work impact the field and change the conversation?

Many of us are familiar with the concept of an elevator pitch but still struggle to apply these ideas in practice. Here is an exercise that illustrates this point:

Exercise 1.1 The Elevator Pitch

1. This exercise can be accomplished with just two participants, although a larger group provides richer feedback.
2. Have the participants write down their elevator pitches based on the outline above and then try to commit it to memory. It is helpful to agree on the audience for this pitch (the general public, a conference presentation, your advisor, etc.).
3. When all participants have created an elevator pitch, designate a leader who will control the timing of the exercise.
4. Divide into groups of two or three, preferably with people you do not know well, and determine who will speak first.
5. On the leader's signal, the first speaker will share their prepared two-to-three-sentence elevator pitch. They will have 30 s to accomplish this task.
6. When 30 s are over, the leader will call time and ask the next group member to present their elevator pitch. Continue in this way until all members of your small group have had a chance to pitch their research.
7. If there are enough participants, switch groups and find a new collection of people. Repeat the exercise under the leader's guidance. Again, take only 30 s per person.
8. Gather the entire group into a circle. The leader will select someone to identify the first person they met during the exercise and ask them to describe what that person does, providing as many details as possible.

9. Check back with the person who provided the research pitch to verify the accuracy of this information. If there are few details, ask the person who pitched their research what is missing that might distinguish them in their field. If the interviewer remembers a rich collection of details, have the group analyze why these details were so memorable.

   This is the teaching moment. When participants see where their elevator pitch succeeds or fails to communicate or inspire, they can begin to determine why and work to make improvements. It is this personal "aha!" that allows a participant to really see what is wrong with the pitch they created. An elevator pitch may look good on paper, but if it does not resonate with the audience, it fails to do the job. We find that the most memorable elevator pitches come in the form of a great story.

10. You don't need to review every participant's experience. Once everyone understands the point of the exercise—*your message may not be getting as much information across as you thought*—have them return to their first small group and exchange what each remembers about the other's elevator pitch. This feedback will be important for revising the pitch.

11. Participants should now return to the written version of their pitch and revise it based on what they have just learned.

    The elevator pitch is a touchstone for research presentations. It encapsulates key information that a researcher can reference regularly in professional conversations. It is its own tiny story. Take the time to perfect your elevator pitch. It will be time well spent (Fig. 1.1).

Fig. 1.1 Working on the elevator pitch

We find that the best elevator pitches come in the form of a story. This strategy will extend to your research presentation. Communications consultant Garr Reynolds (2012) suggests that when creating a presentation, one should consider the techniques of a documentary film. A documentary film intends to educate the viewer on a chosen subject using a cinematic frame to create a compelling story. The storytelling process does not diminish the relevance of the material, but rather engages the audience intellectually and emotionally with the subject at hand. Because so many people watch film and television, your audience will already be habituated to experience your story in a specific way. In this book, we will teach you how to appropriate the viewing habits of your audience and turn them into an advantage in the construction of your research story.

To illustrate our approach, we will focus on the 10 min research talk typically presented at conferences. There are several types of research talks you may need to prepare in addition to our 10 min talk example: the 30 s elevator pitch, the 2 min research summary, the 20 min conference talk, and the 60 min colloquium presentation. The 10 min talk usually has all of the elements of a longer talk but is short enough to be conveniently workshopped while learning storytelling techniques. Once you have mastered the 10 min talk, it is easier to see how to create an hour-long talk since this is just a collection of shorter talks with a common theme. The 10 min talk also creates the opportunity to prepare a 30 s or 2 min summary because it forces the storyteller to focus on the core ideas they would like to convey.

We have divided this book into three main sections. We begin by addressing the content of your presentation and showing you how to improve the story you want to tell. Once you have created a compelling narrative, we will teach you how to include performance techniques to better present your story for an audience. After you have mastered the content and polished your performance, the third section will offer advice regarding logistical elements that often impede speakers during the speaking event. The chapters are constructed as learning modules that can be rearranged. For example, you may want to interweave the first two sections of the book. We encourage you to experiment and adapt this manual to meet your needs.

You can apply many of the ideas in this book on your own. We provide exercises at the end of each section that you can practice alone or in small groups. Each of our exercises have been field-tested in our science communication workshops and designed to specifically address issues encountered in research talks. Importantly, it will not be enough to just read the exercise descriptions; you will only learn the material by applying the concepts *through practice*. You will make mistakes. It will take time to become a master story-teller. But once you have learned how to create a dynamic live performance of your research story, you will find that this professional obligation is no longer something to dread, and you may even come to enjoy presenting your research in public.

Part I

The Story

2

Crafting Your Core Message

Common criticism in some academic circles suggests that making a presentation more polished and including carefully crafted images will hide problems in the research. In our experience, the opposite often happens. A clear talk makes it easier to understand the good ideas, and illustrates where arguments may be flawed or illogical. A clear talk also gives the audience a chance to engage with the research and the researcher in a meaningful way. Moreover, the process of clarifying ideas in a research presentation often helps the *presenter* see how to improve their *own* research.

The clarification process starts with the *Core Message*. A core message is a statement, between one and three sentences long, that encapsulates the primary idea you want to convey to the audience. In addition to being succinct, clear, and jargon-free, this message also needs to communicate why your audience should care about your findings. While the elevator pitch gives context to your research in general, the core message communicates specifically what you have found or achieved with this research project and why it is important. It is the message that you would like your audience to take home. The core message will provide the foundation for your research talk. It serves as the first step and most critical element in your story creation.

Guidelines for Creating a Core Message for a Research Talk

Keep It Simple
Your research project probably covers a lot of territory. You will naturally want to include many elements. Resist this impulse. Focus your core message on the

© Springer Nature Switzerland AG 2019
I. Cohen, M. Dreyer-Lude, *Finding Your Research Voice*,
https://doi.org/10.1007/978-3-030-31520-7_2

most important concept you are trying to convey in your talk. Too much information is hard on your audience. A focused idea will help them process what you want them to learn. Find the essence of your research (Heath & Heath, 2007). What is the one thing you want them to remember (Reynolds, 2012)?

Watch Out for the Curse of Knowledge

The curse of knowledge is the process of forgetting what it was like before you knew what you know now (Heath & Heath, 2007). This obstacle may manifest itself in a couple of ways, and we suggest the following tips to combat it:

1. *Forgetting to connect the dots*: You fail to explain critical components of your research because you assume everyone "already knows that." For example, it does not make any sense to tell someone how to play a diminished third on the piano if they do not understand musical scales. Determining the appropriate level of knowledge for your audience can be challenging and may require feedback. It helps to know the background of your audience. Our experience is that presenters are often speaking at a level that is too high, even at discipline-focused conferences. After all, everyone is somewhat of a layperson outside of their own particular research specialty.
2. *Cut the fat*: You include irrelevant information, assuming everyone would want to know how this or that process works or was developed. This extra information, though fascinating to you, may distract from the main narrative. You must be brutal when you edit and be willing to kill those beautiful side stories that took you years to create. Meyers and Nix (2011) insist that the most common complaints from a conference audience are: "(1) Too much information. (2) Not relevant. (3) No point." (p. 44). Do not assume your audience will know everything about your research, but do not compensate for this lack of knowledge by including everything you know. Keep your core message lean and clear.

Avoid Jargon Whenever Possible

Jargon has its place. When speaking with your advisor or other experts in your field, jargon is a useful shorthand for getting a lot of information across quickly. In a talk to a broader audience, however, jargon stands in the way of clarity. You may happen to be an expert on the scallop theorem and reciprocal motion in low Reynolds number flow, but unless you specialize in fluid mechanics, that statement is incomprehensible. Instead, it is easier for the audience to understand that different swimming strategies are necessary when a pool is filled with honey instead of water. You may think that jargon makes you sound smart, but it requires more processing time for the listener. The best presenters are able to say things simply without dumbing down their presentations.

Exercise 2.1 Beginning to Craft the Core Message

1. Look at all of the material you want to communicate and reduce it to the essential idea. You should be able to articulate this idea in a few short sentences. It will take multiple attempts to reduce your research to a core idea. Do not worry about getting it right the first time, just write many possible versions trying to follow the guidelines listed above.
2. Once you have something you like, try to reduce any use of jargon and make the tone of your message relatable. A good core message is something that someone outside of your specific discipline can understand.
3. Make sure you have included why we should care. Do not take for granted that your research is important, help us understand why and how.

Working on your core message can be challenging. Often we find ourselves in circular ruts. The following improvisation exercise will help you further simplify your crafted core message by forcing you to utilize the right side of your brain. This is best done in a group of three or more. *We recommend you use a camera to record the entire exercise.* You will forget what you say, and some of your most creative material may be lost as a result. Record your work and play it back. Watching yourself may be difficult, but there is no faster way to adjust your presentation, and it is worth the temporary discomfort.

Exercise 2.2 Refining the Core Message with DING!

Ding! is an improvisation game that teaches you how to instantly create a new solution to an old paradigm.

Phase 1: Regular DING!

1. Two people come to the front of the room.
2. The audience helps set up the exercise by providing important context for the game. They need to choose:

   Location: (the beach, a nightclub, on the moon, at the car shop)
   Relationship: (siblings, lovers, a rival from high school, employee/boss)

3. The actors now begin to improvise a scene. The following important rules will help this go well.

   Choose something active to do. If you are on the beach with a sibling, ask him or her to put sunscreen on your back. If you are on the moon with your lover, discover that you have a terrible itch in the middle of your foot. If you are at the car shop with a rival, offer to give him or her a ride home. This initial choice will lead you somewhere. Be as imaginative as possible.

Say yes to every suggestion. The golden rule in improvisation is "Yes/And." That means that if someone looks at you and says, "Wow! I didn't know you were pregnant!" you cannot say "No, I'm not" (even if you are male). You might say "Yes I am. And you're the father!" or "Yes, isn't it amazing? Septuplets!" "Yes" always leads to a new and interesting idea. "No" shuts the action down.

4. As the scene unfolds and events begin to happen, the audience can shout out at any point, "*Ding!*" Ding! functions like a game show buzzer. The actor who has been "dinged" must pause, wind the action back to the beginning of the last sentence, and then say the sentence again with a new ending. It might work like this: You and your sibling are on the beach. She comments on the nice tan you have. You say, "Yes, isn't it great? It only cost me $9.95." Someone in the audience shouts "Ding!" You wind the action back and say, "Yes, isn't it great? It goes well with my bathing suit." Ding! "Yes, isn't it great? It guarantees skin cancer by the time I'm 40." Or perhaps your partner says, "Boy this sun is hot, I could sure use a drink." Ding! "Boy this sun is hot, I could sure use a trip to Alaska." Ding! "Boy this sun is hot. I could sure use this opportunity to discuss the merits of proper retirement planning." Moderators should try to encourage participants to explore a variety of possible endings. For example, if answers fall into a similar category (food, locations, etc.) ding them again until they find something new.

5. Play a scene for up to three minutes. When the players catch on and have the idea, thank them and offer applause. Then switch out until everyone has had the chance to try (Fig. 2.1).

Fig. 2.1 Ding!

Phase 2: Research DING!
This is the part you will want to record on your phone. It is easy to forget what was just said, and a recording will allow the researcher to rewind and select great moments for a future research talk (Fig. 2.2).

1. Choose a moderator to guide the process.
2. The moderator asks one person to come to the front of the room armed with a memorized core message.
3. This participant should recite the core message to the audience as crafted. Allow this participant to get all the way through the constructed core message.
4. Have the participant say the core message again, allowing audience members to Ding! them whenever they encounter a section that is difficult to understand or is full of jargon. It is important that the audience not be shy about "dinging" the person on stage. The quick-fire process of getting dinged during this part of the exercise is what helps release creative versions of the core message. On the other hand, let the participant catch their breath or finish their thought between dings.
5. If the participant is still struggling to simplify the core message, have them "tell it to a fourth grader." This helps eliminate jargon and the curse of knowledge.
6. Finally, have the researcher start the core message one more time, trying to stay within the one-to-three-sentence limitation.
    We find that through this exercise the researcher often creates golden phrases that beautifully capture the central idea of their research. The mod-

Fig. 2.2   Recording the Ding! exercise

erator and the audience should be sure to point these out. These phrases can be fleeting, which is why it is important to record the exercise for later playback.

7. Make sure everyone in the group participates in this process.
8. On your own, watch the recorded versions of your Ding! exercise and extract ideas that will be useful to you.
9. Rewrite your core message with this new information.

Here is an example of how we used Ding! in a workshop we conducted:

Dario's Ding!

First attempt at the core message:
Membrane proteins are the high hanging fruit of drug design. Membrane proteins are hard to express, making the research very prohibitive. We have engineered a way in which membrane proteins can be expressed and solutized in a detergent-free manner making the research more accessible.

(The message is headed in the right direction but is suffering from many of the ailments listed above.)

DING—to a 4th grader:

In our bodies, we have cells. We are made out of millions of these cells. And although we have eyes and ears and we can see what's going on, our cells are the ultimate judge of what is going on. So, if you have a fever, your cells realize they have a fever. You're going to go to the doctor, and the doctor is going to give you medicine that will tell the cells to get better. The medicine needs to be precise and match your disease or the thing that is making you sick. For a doctor to understand what is the exact medicine you need for your sickness he needs to understand little proteins that live in your cells. Those proteins are very hard to understand. The doctors need new tools to understand them better and that's the research.

(Although this version rambles, Dario is breaking down his research in new and interesting ways.)

DING! with interruptions:

Ding! #1   *Our research engineers chaperones to enable membrane research for drug designers.* (clearer but still vague)

Ding! #2   *Our research eliminates the need for detergents in membrane protein research.* (important but not the critical point)

Ding! #3   *Our research will enable membrane protein research in drug design in a cheaper way.* (another version of #2)

Ding! #4   *The technology that we have engineered allows membrane proteins to maintain their function even in an alien environment, making research more accessible.* (Much clearer and captures both. Still needs a bit more work but well on its way.)

Final Core Message:
Our research enables membrane proteins to maintain their function even in an alien environment, which will allow drug designers to access them less expensively, opening up the field for new drug development.

Once you have refined your core message, it is important to test it to determine if others understand what you are trying to say. This is stage three of core message construction.

Exercise 2.3 Workshopping Your Core Message

1. Have a volunteer share their core message.
2. Discuss the draft collectively and identify what is working and what needs improvement.
3. Check the core message for jargon. Have the volunteer deliver the core message again and ask audience participants to raise their hands if they do not understand a particular concept or idea. Try to brainstorm together to find a clearer phrase or an alternative way to express the same idea more simply.
4. Repeat the process allowing all participants to receive feedback.
5. Incorporate the feedback you received into your core message.
   It is uncommon for your core message to expand to include all the ideas you have been given. Pick and choose what works for you. The final stage of this process is to reduce the core statement to its essence so that it can be stated in one to three sentences.

Core Message Example 1—Draft 1:
Using confocal microscopy, we have imaged the 3D location of individual colloidal particles and using probabilistic methods estimated the collision frequency and directionality of particle collisions to extract the forces on the single particle scale.

Now we'll apply the concepts listed above:

- **Keep it simple** → too complex
- **Watch out for the Curse of Knowledge** → this is present
- **Avoid jargon** → lots of jargon here

Core Message Example—Draft 7 (having gone through Ding! and an extensive editing process):
We have developed a method for turning our microscope into a local pressure gauge.

Applying the concepts again:

- **Keep it simple** → Yes. Word count dropped from 35 to 12.
- **Watch out for the Curse of Knowledge** → None.
- **Avoid jargon** → None. We can all understand what a microscope and a pressure gauge are.

Core Message Example 2—Draft 1:
The melodies inherent in a foreign language performance contain the potential to trigger psychocognitive empathy through emotional contagion and mirror neuron processing, broadening the potential social impact of multilingual performance.

Now we'll apply the concepts listed above:

- **Keep it simple** → both long and complex
- **Watch out for the Curse of Knowledge** → it is not clear to the layperson what this means
- **Avoid jargon** → lots of jargon here

Core Message Example—**Draft 7** (having gone through Ding! and an extensive editing process):
The music in language can trigger an empathic response even when we do not understand the words, and those feelings can change our minds about people different from us.

Applying the concepts again:

- **Keep it simple** → Yes. Character count dropped from 238 to 167.
- **Watch out for the Curse of Knowledge** → None.
- **Avoid jargon** → None. (How and when this idea might be applied will require more explanation.)

This process may seem tedious and unnecessarily time consuming. Remember that the core message serves as the touchstone for your entire research talk. Refining it is worth the effort. Once you have created the core message for your research talk, you are ready to put it into action. The following exercise demonstrates how this works.

Exercise 2.4 Using the Core Message to Focus Your Research Talk

1. Review your talk slide by slide. For each slide, construct a summary sentence that captures the main point you are trying to convey. You can also perform this exercise for each image or graph. Try to determine the main point of the slide or image and what the audience should see to understand that point. Write this idea in your summary sentence.
2. Go through these summary sentences and remove any slides that are not relevant to the core message you have constructed for your talk (Fig. 2.3). Your goal is to determine the fastest way to get from the beginning to the end of your talk.

   - Get rid of the title slide.
   - Get rid of the outline slide.
   - Keep the slides explaining the main research problem.
   - Keep the slides explaining a crucial technique you are using to solve this problem.

- Remove any side stories. This includes most of the background material that is not your own research. You must figure out how to retain the minimal amount of background information without having it dominate your talk.
- Remove unnecessary math calculations.
- Remove slides explaining a cool simulation that is not strictly necessary for the main message.
- Keep slides that summarize results.
- Keep slides that help the audience to understand the results.

3. Review the remaining slides and determine if you have a complete narrative. If there are holes or gaps, this is the time to construct slides that would fill these gaps.
4. Be brutal, and make sure that the new slides are still absolutely necessary to complete the main idea identified in your core statement.
5. It may be useful go through this exercise as a group using the talks of one or two volunteers. For each talk, have the listeners identify the simple line of reasoning that is being presented. If there are gaps, fill them. If something is not in order, move it into the right place.
6. If there is not sufficient time to go through everyone's talk, try breaking up into groups of two or three, and have each subgroup repeat the exercise with each person's trimmed talk. Peer feedback from researchers outside the discipline is particularly helpful to reveal misinterpretations as they share what they interpret from each image or graph.

Fig. 2.3 Working on the core message

If you would like advice on how to shape the aesthetics and effectiveness of your slides, images, and data representations, there are multiple resources available. Be sure to seek guidance that is appropriate to your discipline. In our experience, slides often contain too much information. Try to use fewer words, fewer equations, and only one or two images. Do not worry about blank spaces. Make sure each slide visually conveys a single idea. Here are some resources for working on the aesthetics and layout of your slide presentation:

- *Dazzle'em with style: The art of oral scientific presentation* (Anholt, 2010).
- *Creating Effective Slides: Design, Use, and Construction in Science* (Doumont, 2013).
- *How to Avoid Death by PowerPoint* (Phillips, 2014).
- *Slide:ology: The art and science of creating great presentations* (Duarte, 2008).
- *Storytelling with data: A data visualization guide for business professionals* (Knaflic, 2015).
- *Presentation Zen: Simple ideas on presentation design and delivery* (Reynolds, 2012).
- *Visual strategies: A practical guide to graphics for scientists and engineers* (Frankel & DePace, 2012).

Crafting a core message may be the most difficult part of designing your talk. Once you have identified your core message, however, the rest of the talk becomes much easier to create. A 10–20 min talk usually has one core message. An hour-long talk still has one core message but may have two or three interrelated themes framed by this overarching idea. The core message is the foundation for your presentation and determining a good core message is well worth the effort.

3

Shaping the Dramatic Arc

The shape of your story is a critical component in effectively transmitting your core message and inspiring interest in your research. Many researchers make the mistake of assuming that presenting the facts in a logical, sequential manner will be the most effective way to convey the content of their work and the professionalism of their approach. The problem is your audience may be asleep before you communicate your essential finding and may remember little of what you worked so hard to tell them. When someone steps into a room (a theater, a lecture hall, a conference auditorium) and sits down to listen to a speaker who wants to engage their attention for a period of time, they unconsciously begin to expect certain storytelling events to happen. They expect the speaker to pique their curiosity, transport them to another place, and keep them wondering how the story will end (Heath & Heath, 2007).

Why does this matter? Scientists come to a conference or research presentation to hear about science and just want the facts, right? Yes and no. The science must be presented credibly, but facts are dull and hard to remember. People viewing a presentation seek an emotional or intellectual connection. Gallo called this "brain-to-brain coupling" (2014), and it is this coupling that stimulates your audience to pay attention. A good story introduces situations that pose compelling questions. Heath and Heath named this question-posing strategy "the gap theory," because a good speaker needs to open a gap in a body of knowledge before closing it with facts. In a research talk, opening a gap creates the need to know something about your topic. The gap

Electronic supplementary material The online version of this chapter (https://doi.org/10.1007/978-3-030-31520-7_3) contains supplementary material, which is available to authorized users. The videos can be accessed by scanning the related images with the SN More Media App.

is a problem that requires knowledge to solve. Opening gaps hooks the audience and keeps them engaged. A list of facts, on the other hand, closes gaps. With closed gaps there is no active cognition required. Facts do not inspire curiosity; facts do not make one wonder; a list of facts fails to create mystery. A good story contains open gaps that entice your audience into believing that you have something extraordinary to share, something that matters, or something that will change their thinking. Once you have their interest, you can close the gaps you have created with the facts you discovered during your research. Starting with the facts in a research talk shuts down the opportunity for discovery. Consider saving your facts for the best storytelling moment in your presentation, allowing your facts to serve as the answer to the mystery you have positioned with the rest of your talk.

We are hard-wired to absorb information in narrative form (Reynolds, 2012). Paul Zak insists that the process of becoming absorbed in a story is biochemical:

> Stories are powerful because they transport us into other people's worlds but in doing that it changes the way our brains work, potentially changing our brain chemistry…Dramatic stories cause us to care about others. When we hear them, chemicals are released in our brains spurring us to action (2012).

Researchers have tested the impact of storytelling on the retention of facts. Students at Stanford University were asked to remember the content of presentations that included both facts and stories. Only 5% of them remembered the facts, but 63% of them remembered the stories (Heath & Heath, 2007). If you learn to convert your research into a compelling story, you have a statistically higher probability that you will engage your audience, and that they will remember your presentation.

Some stories are best told as a linear progression and some are not. A murder mystery, for example, rarely starts with the murder. Instead, it starts in the middle, when the detective arrives and uses flash backs to assemble the plot. If the story was told in sequence you would already know who the murderer is after the first scene! You will need to determine the best way to tell your story, keeping in mind that the order in which you tell it can make a big difference.

Structure

By becoming conscious of how a story is constructed, you can begin to control the order of events in your research story for maximum impact. It will therefore be helpful to review the structure of a story, and to define some terms before we investigate how to apply them to a research talk. Some basic building blocks of any story are: *exposition, inciting incident, complicating actions, climax,* and *resolution.* These elements can be arranged and rearranged to suit your storytelling needs.

Exposition Exposition is the part of the story containing all of the circumstantial elements your viewer needs to know in order to understand what is going on. Where are we? When is it? Who are the players? What are their relationships? What events have occurred prior to our story that we must know in order to enter the narrative smoothly? In your research talk, the audience will want you to provide background so that they can understand some basic information. What question are you researching? Why is it important? What have other researchers done? Exposition can be tricky. It is critical that your audience have enough information to connect to the events, but too much information can become confusing and dull (McKee, 1997). The other delicate aspect of exposition is when to introduce it. Traditionally, it comes at the beginning of the story, but many stories resist this standard form. Expositional elements can, for example, be woven throughout a story, providing the audience with just enough background information for the next narrative event. There are several strategies for including background information in your research talk. It is important to set the scene, but be sure to carefully calibrate this component of your story—just enough but not too much.

Inciting Incident The inciting incident is the one event without which there would be no story. It is often the moment that sets up the conflict in the plot (Thomas, 2013). In your research talk, an inciting incident could be the moment that propelled you to investigate the subject of your research. What happened that started you thinking that this should be your next research topic? Did you have a new idea or develop a new technique that could shed light on some aspect of your project or allow you to tackle the research problem in an innovative way? Something sparked your interest in this particular research topic. Identify what that something was. Alternatively, the inciting incident could have been a failed project. What did that project teach you, and how did it allow you to change direction, or make progress on the topic you are investigating now?

Complicating Actions Probably the most critical aspect of storytelling is building tension through conflict. This is where complicating actions appear. In most stories, one finds a protagonist pursuing a specific goal. If the hero or heroine grabs the prize without encountering any obstacles, the story is wholly unsatisfying. We are engaged with a story when it surprises us—when our expectations are thwarted or frustrated (Reynolds, 2012). When watching or listening to a story, we expect our heroine to encounter challenges, disappointments, and frustrations, and we want the protagonist to overcome them. Living vicariously through this process is central to dramatic catharsis, and one of the things that makes good storytelling so satisfying.

In a research talk this series of hurdles may come in the form of experiments that you ran and are trying to interpret or understand. You run the first experiment and get a particular result. Maybe this result is confusing, which raises a question that drives you to conduct a second experiment. Maybe that second experiment gives a result that solves part of the problem but suggests a third experiment to understand things more fully. Finally, the third experiment may yield a result that sets up the final conclusion. Presenting the conclusion or answer to your research problem will be more satisfying for your audience if you share the hurdles you had to overcome to get there.

In many stories, the protagonist will encounter one or more antagonists. Everyone loves a good villain, and villains are often the main source of conflict. In your research story, you may find that you or the solution to your research problem serve as the protagonists. Few research investigations run smoothly. There are almost always unexpected events along the way. Perhaps these events are your antagonists. They may become your complicating actions. You may be tempted to disguise or hide challenges you encountered in your research process (a particularly grueling experiment, unexpected results, or the need to build a new apparatus) in order to save face. Resist that impulse. Obstacles, unexpected findings, or even failures along the way are storytelling gold. They increase dramatic conflict and help set up the climax. By sharing them with the audience, you can build tension in your story, and your audience will care more about how the story ends. The key to maximizing the storytelling value of obstacles is to brag about conquering them rather than complain about having to work so hard to overcome them.

Climax Once you have established your complicating actions, you are ready to reveal the climax of your story. Your inciting incident introduced the problem, and your complicating actions developed conflict and tension. The pressure on your research problem should be at its greatest point when you reach the climax in your research talk. The climax happens the moment before the resolution. It is the moment of greatest emotional intensity in a story (Thomas, 2013), and brings about a significant change in the problem at the heart of your narrative (McKee, 1997). If you have set your story up properly, the climax will feel inevitable. This has to happen; this must be resolved. After the build in tension created by the complicating actions, the climax-to-resolution transition provides the long-awaited emotional catharsis for your audience. Granted, your audience may not be moved to tears at a research conference, but if you have hooked them with the initial part of your story, the climax is where you can provide a satisfying payoff for their attention and time. This climax will probably center around your primary discovery or the new piece of knowledge that emerged from your battle with the problem introduced in the inciting incident.

Resolution The resolution is the final piece in the storytelling structure. Sometimes referred to as falling action or denouement, this is where you wrap up any loose ends in your story. This is the place in your talk where you present your research breakthrough and take-home message. It is also an opportunity to zoom out and let the audience know about the potential impact of your research findings on the broader community.

It may be helpful to see some examples of how this story analysis applies. First, we will apply the terms to a television program so that definitions are clear, and then we will demonstrate how to apply them to a research talk.

Example 1

House—Season 1, Episode 2 "Paternity", first broadcast on November 23, 2004. (Watch it first if your resources permit.)

Overview: A 16-year-old male comes to the hospital complaining of double vision and night terrors after being hit in the head by a lacrosse stick. Doctor House is dismissive until he notices a jerk in the boy's foot. After a near-fatal hallucination and several faulty diagnoses, House is mystified until he learns the boy's true paternity.

Story outline

Exposition	(Laced throughout). A high school lacrosse player with two loving parents suffers from night terrors and double vision.
Inciting incident	A young man is hit in the head with a lacrosse stick during a game and is rushed to the emergency room. (If he had not been injured this story would not need to be told.)
Complicating actions	1. In the emergency room, House sees an unusual knee jerk, which destroys his previous diagnoses of this young man's condition. 2. They discover a broken brain vessel and repair it. 3. The young man begins to hallucinate and nearly walks off a roof edge. 4. The new diagnosis is neurosyphillis. They inject penicillin directly into his brain. 5. The young man has auditory hallucinations, destroying the previous diagnosis. 6. The young man's parents argue with House over the lack of progress. House steals their coffee cups and runs a DNA test. 7. We discover that the young man was adopted. They run a test for the measles virus. He tests positive.
Climax	As the young man moves closer to death, the doctors rush him into surgery to correct the problem.
Resolution	Brain surgery goes well. He recovers full function. His parents are no longer angry with House.

Example 2 *Itai Cohen Group's Complex Matter Lab: The flight of the fruit fly.* Presented at 71st Annual Meeting of the American Physical Society Division of Fluid Dynamics on November 18, 2018. Atlanta, Georgia (Cohen, 2019; Video 1 in Supplementary Material).

Overview: Insect flight is aerodynamically unstable. Flies have developed strategies for altering their wing motions to allow them to stabilize their flight. The Cohen group investigates these strategies by gluing magnets onto the flies and applying magnetic field pulses to perturb the flies in midair. By analyzing the resulting wing motions the group has determined that insects use a proportional integral controller to adjust these wing motions and maintain their body pitch orientation. These controllers are very similar to those used to set the speed of a car under cruise control or set the temperature of a room. To investigate the neural origins of this control circuit, they use genetically modified flies where individual neurons in the circuit governing these controllers can be manipulated. Finally, they discuss why flies may have evolved this control apparatus and the impact that evolving flight control had on the ecology of our planet.

Story outline

Exposition (background information to set up the story)	Itai introduces the hero of the story (the fruit fly) and explains basic aspects of control through a demonstration: balancing a pipe on his fingertips. (1:00–2:17)
Inciting incident (introduction of the problem that drives the story)	To fly stably, a fruit fly must do this type of balancing at an incredibly fast timescale. How does it do this? (2:17–2:47)
Complicating actions (Chapters of the investigation. Each chapter builds on the other and increases the complexity and interest of the research topic)	Chapter 1: How are fruit flies producing forward thrusts and turns? (2:50–16:00)—also contains some of the exposition. Chapter 2: How does the fly maintain stable flight? (16:00–22:01) Chapter 3: The relevance of a gyroscope to stable flight, genetic manipulation of the control circuit, and the evolutionary value of halteres. (22:01–30: 31)
Climaxes	Chapter 1: Demonstration of turns. (12:00–13:00) Chapter 2: Putting the pieces together. (19:30–20:00) Chapter 3: Comparison of pitch perturbation experiments in normal versus b1 silenced flies. (24:20–25:00) Chapter 3: Glue a dandelion fiber onto the fly and stability is restored. (29:25–29:30)
Resolution	The evolution of halteres and strategies for stable flight has had a massive impact on the ecology of our planet. (30:40–31:07)

Exercise 3.1 Identifying the Elements of Your Dramatic Arc

1. *Give your talk a title.*
   Both of the examples above have titles that identify the talk. The television episode is entitled "Paternity." The research lecture's title is "The Flight of the Fruit Fly." How would you title your talk? Keep it simple but clear.
2. *Find the primary ideas that will help support your core message.*
   Meyers and Nix (2011) recommended identifying the three big things you want to introduce in your talk. The number of items may depend on the length of the talk. In a 10 min talk, choose one story to support your core message. In a 20 min talk, choose two or three. In an hour-long talk, narrow it down to four or five stories that support the big idea that you want to convey. Write down your main stories.
3. *Create your dramatic arc.*
   Using the chart below, begin to identify the components of your research story. Do not worry about shape or length at this point, just list what you believe belongs in each box. We recommend beginning with the climax/es, as everything builds to that big idea or discovery. The most challenging box will be your complicating actions. You need to take us on a journey to your big discovery without including everything that happened along the way. Be selective (Fig. 3.1).

Fig. 3.1 Working on the dramatic arc

Exercise chart

Exposition
  (The setup. What do we need to know to understand your story?)
Inciting incident
  (What was the moment when this all started? What happened?)
Complicating actions
  (What sequence of events led to the climax of the story? They should build
    tension as they unfold)
Climax
  (What was the big moment? The crowning event? The aha!?)
Resolution
  (Tie up all loose ends here. Fill in the blanks. Or suggest future research that will
    emerge from this project)

Shape

"Like every book, movie, opera, or play ever written, your talk has three parts: a beginning, a middle, and an end" (Meyers & Nix, 2011, p. 49). The "beginning" in the dramatic arc covers the exposition and the inciting incident. Complicating actions and the climax comprise the middle, and the resolution provides the ending of your story. Like an hourglass, the beginning and end are broad, while the middle focuses on the details of your particular study. Although this may all seem obvious from the descriptions above, what may not be obvious is that these components can be rearranged to provide the biggest impact for your particular story. You have an obligation to present credible, compelling research using the tools of the profession. But there are no rules that compel you to begin with the beginning or end with the end. Television programs like *Breaking Bad* often tease the audience by beginning an episode with a short clip of an important event (usually a complicating action) that happens later in the story. This initial bit of narrative is out of context and therefore mysterious. The viewer is left to wonder when that particular moment will be revealed and explained during the story. A number of novels begin with the end of the tale, and then take us back in time to show us how we got there. Serialized television shows (soap operas being the most notorious) often end an episode with a new unresolved piece of information to encourage the viewer to "tune-in next time." To make your research story compelling, you must find the shape that best suits the story you want to tell. You will need to determine what shape will best serve the experience you want to create for your audience.

The following improvisation exercise will help you see the value in considering nontraditional structures for your research story.

Exercise 3.2 Story Parts

1. Gather in groups of four or five. Have each member of the group tell a short story about something memorable that happened to them. This could be an old memory (when you fell out of the tree at age seven and broke your arm) or a recent memory (when your mother showed up and surprised you for your birthday).
2. Choose one of the stories in your group to serve as your material. You will not be using everyone's story so the story you remember best might be a good candidate. Make sure it has a beginning, a middle, and an end.
3. You will be telling this story to the audience as though it were your own. Assign one section of the story to each member of the group. These sections should roughly relate to the Dramatic Arc. Someone will provide the exposition or setup, another will introduce the problem (the inciting incident), one member will relate complicating actions, another the climax, and finally someone will resolve the story. If necessary, group members can be in charge of more than one section of the story arc.
4. Stand in a line on the stage and tell your story to the audience, one at a time, using the first person. When it is your turn, tell it in the order you heard it.
5. Now keep your roles in the story, but rearrange your positions and tell the story again, adjusting the detail of the narrative to make sense of the story. The person telling the end may start or someone who told a portion of the middle. This way your story will start in a new place. Just be sure that you do not follow the same pattern that you used the first time. While this way of telling may not be chronological (in time), it may have greater impact and keep the audience more engaged.

Example:
(One member of your group tells the following story.)

When I was 10 years old, my mom took me on a trip to Uganda. At one point we went on safari and we drove through this big park full of animals. It was really cool. We stopped the car to wait for parking, and a baboon hopped up onto the car, then the roof, then through the sunroof and down into the car and it started to attack me. My mother was having none of it. She grabbed the pillow I had been using to nap on and started beating that baboon. It just hissed at her. So she got really mad and began hollering and beating and scared the heck out of that baboon. She chased it back up and out of the sun roof, off of the car, and out into the park. I'll never forget that. That baboon was almost as big as I was, and it had gigantic yellow teeth. Didn't matter. It was no match for my mom.

(Each person in the group now takes a piece of that story and tells it to the bigger group using as though it were their own.)

Person 1 (exposition): *When I was 10 years old, my mom took me on a trip to Uganda. At one point we went on safari and we drove through this big park full of animals. It was really cool.*

Person 2 (inciting incident): *We stopped the car to wait for parking, and a baboon hopped up onto the car, then the roof, then through the sunroof and down into the car and it started to attack me.*

Person 3 (complicating actions): *My mother was having none of it. She grabbed the pillow I had been using to nap on and started beating that baboon. It just hissed at her. So she got really mad and began hollering and beating and scared the heck out of that baboon.*

Person 4 (climax): *She chased it back up and out of the sunroof, off of the car, and into the park.*

Person 5 (resolution): *I'll never forget that. That baboon was almost as big as I was, and it had gigantic yellow teeth. Didn't matter. It was no match for my mom.*

(Now rearrange the order and tell the story again, making minor adjustments in the narrative to make sense of the whole.)

Person 5 (resolution): *I'll never forget [the time I was staring face-to-face with a] baboon [that was] almost as big as I was. [I]t had gigantic yellow teeth. Didn't matter [how big it was]. It was no match for my mom.*

Person 2 (inciting incident): *[What happened was] we [had] stopped the car to wait for parking, and a baboon [had] hopped up onto the car, then the roof, then through the sunroof and down into the car and it started to attack me.*

Person 3 (complicating actions): *My mother was having none of it. She grabbed the pillow I had been using to nap on and started beating that baboon. It just hissed at her. So she got really mad and began hollering and beating and scared the heck out of that baboon.*

Person 4 (climax): *She chased it back up and out of the sun roof, off of the car, and into the park.*

Person 1 (exposition): *[This all happened] when I was 10 years old [when] my mom took me on a trip to Uganda. [We were on] safari and we drove through this big park full of animals. It was really cool. [Except for the baboon.]*

This exercise helps demonstrate the value in rearranging the parts of your story. You can create tension, build suspense, or generate surprise through the simple act of choosing not to begin with the beginning. Now apply these ideas to your research talk.

Exercise 3.3 Determining the Shape of Your Story

In a group or on your own, consider some alternatives to a traditional narrative plot. It might be helpful to storyboard your talk (Reynolds, 2012). Storyboarding is a technique used by filmmakers to arrange the components of a screenplay so they can more easily organize each shot. Usually a storyboard is a graphic representation (a thumbnail sketch) of each story moment laid out in sequence.

In PowerPoint, it is convenient to storyboard using the slide sorter (see Fig. 3.2). Your storyboard can take any form that is useful to you. Think of the components of your dramatic arc as pieces of a puzzle that you can arrange in a variety of configurations.

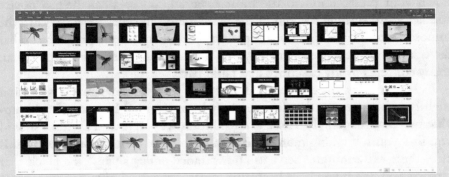

Fig. 3.2 Storyboarding with PowerPoint

Looking at your dramatic arc, what would happen if:

- You began with the end?
- You began with one of your complicating actions?
- You constructed a murder mystery from your story?
- You created an action adventure with your story?
- You wrapped your story around the experiences of your hero? Or your villain?
- You modeled your story shape after a story you know that is particularly compelling?

   Notice how the focus of the story shifts when you start at different places in the linear narrative. Now determine the best shape of your story by arranging the components of your arc to emphasize your core statement or to generate the greatest dramatic effect.

Length

It is important that you monitor the length of your talk. A talk that goes on longer than it should can undermine the overall impact of a great presentation. If a speaker goes over their allotted time, the audience gets restless, the next speaker starts to panic, and the session moderator may become annoyed. Determining how long your presentation will take is one of the most challenging aspects of creating a talk. If you do not include enough information, your audience will feel cheated. If you include too much information, they may lose interest, or not grasp which elements of your talk are most critical.

In short talks, it is good to remember that less is almost always more. Paring down your presentation to the most essential concepts can help a novice speaker avoid rushing the end of the presentation in order to stay within the imposed time limits. It is much easier on the audience if the presentation has fewer concepts described concisely at a slow or moderate pace, rather than many concepts overly described at a rapid pace.

In longer talks, we recommend that you *plan* to end your presentation slightly earlier than the deadline you have been given. This approach will provide two advantages: (1) Those extra minutes will serve as a cushion should you find yourself taking more time than normal on certain concepts. In this case, these extra minutes serve as an insurance policy against the clock that restricts you. (2) If you have been able to get through all of your slides, those extra minutes can be used to discuss the big picture context more fully. Either way you win.

If you are presenting an hour-long talk, it is helpful to know that most of the members of your audience will need a break after 18 min (Meyers & Nix, 2011). That is why TED Talks are restricted to an 18 min window. To solve this problem during an hour-long talk, break your narrative down into three smaller sections, each of which builds toward the climax of your story. It is helpful to change the rhythm of the talk at each of these time markers in order to give the audience a break from the steady stream of information. Other options include incorporating historical context, anecdotes, or personal experiences. Master storytellers are able to incorporate some of these elements even in shorter talks.

You will notice in Itai's talk that he incorporates a variety of strategies to keep the audience engaged. He introduces a concept with a demo or an anecdote and saves the complex equations until after he has set up relevance. He divides his talk into three roughly equal sections, each with its own dramatic arc including an interesting beginning and a strong ending. At 20 min (very close to the 18 min recommendation), Itai incorporates a memorable video which simultaneously generates surprise, incorporates humor, and establishes relevance to his research. Through his talk, Itai uses multiple modes of engagement. His slide show is primarily populated by images and serves as a support rather than a repeat of his narrative.

Exercise 3.4 Dividing Your Talk into Sections

Identify where in your talk you can take a short pause or create a shift in your narrative. Natural places to divide a talk are at the conclusion of an experiment or result, a place where you became stuck and needed to reevaluate your research approach, or a shift to the next phase of your research.

1. Use the slide sorter view to look at your talk and identify possible breaking points. Remember, the idea is to experiment with locations where these interludes can be placed.
2. Brainstorm what you can do during these short intermissions to allow the audience to rest for a moment but still keep them engaged.

   Possible ideas:

   * Illustrate the research you've just described with a demonstration.
   * Bring in a sample that you can pass around to the audience.
   * Have the audience conduct a mini-experiment from their seats.
   * Show a short video.
   * Relate what you are discussing to modern day events or technologies.
   * Include a historical interlude that traces the origin of a scientific idea.
   * Tell a short anecdote.

3. Experiment. If you have a great anecdote that is not in the right place, try putting it in a different place. Sometimes you will identify a breaking point but your audience may not need a break. Get feedback from your colleagues regarding whether these small intermissions are working.
4. Test your ideas in front of a live audience. If they work, great. If they sort of work, think about how to improve them. If they don't work, drop them or try a different approach.

Finding small breaks in your story will help your audience stay engaged. Becoming aware of the 18 min window of attention allows you to embed moments of surprise or transition that can strengthen your research story. Breaking your talk into sections can also help you stay within your time limit. You can identify when you might go over the time allotment if you find yourself exceeding your smaller 18 min sections. As you plan the map of your research talk, continue to remember that less is almost always more.

Constructing a dramatic arc for your research talk can be challenging. Once you have found the core ideas in your story, the elements of the dramatic arc will more easily fall into place. Using this structure helps you identify which components of your research story are most essential, and keeps you from including too much information. Running your trimmed dramatic arc by a colleague, especially one who is outside your own specialty, will help you determine whether they are engaged and can easily follow what is going on, or if you have taken out too much. Eventually, a good dramatic arc will allow you to turn your core message into a compelling story, offering a greater possibility that your audience will stay engaged and alert during your entire research presentation.

4

Great Beginnings and Strong Endings

Great Beginnings

First impressions make a difference. The way you begin your research talk can prime the audience positively or negatively for everything that follows. Daniel Kahneman identifies this phenomenon as the halo effect (2011). The halo effect happens when the first impression in an experience colors the way you view everything that happens thereafter. If you begin your research talk well, the audience will be predisposed to listen and enjoy what you have to say. If you begin poorly, you may lose them long before you reveal the most important information. How long do you have to make a positive first impression? One study suggests that you have between 5 and 15 min to impress someone (Frieder, Van Iddekinge, & Raymark, 2016). Another expert states that you have only 2 min before your audience has decided for you or against you (Reynolds, 2012). Meyers and Nix reduce the workable time frame for accomplishing this goal to seven seconds (2011). The exact timing will depend on the context of your presentation, the scheduled length of your talk, and how famous you are. Regardless, all experts agree that a good beginning is critical to a successful presentation. Do not make the mistake of assuming that your audience will listen to you just because you are speaking to them (Meyers & Nix, 2011). You have to earn their attention. You need to captivate their interest. The first words you say and the first thing you do in your research talk are critically important to creating your storytelling hook. You need to immediately engage the audience's attention and create high stakes for your talk, so that those who are watching will want to lean forward to hear the next thing you have to say (Meyers & Nix, 2011).

© Springer Nature Switzerland AG 2019
I. Cohen, M. Dreyer-Lude, *Finding Your Research Voice*,
https://doi.org/10.1007/978-3-030-31520-7_4

It can be challenging to find the right way to begin if you are accustomed to a more traditional approach. It is important that you choose something that matches the message you are trying to convey, but we recommend that you try to go beyond the traditional format. Standard introductions with an opening slide that contains the title, your name, and institution, accompanied by an agenda or outline slide, will signal to your audience that this is a typical talk. Because these standard introductions are less engaging, you may find your audience looking at their phones or studying the conference program while your talk gets going. A great beginning, on the other hand, will engage your audience from the moment you start to speak. Here are a few concrete ideas that may inspire the first few minutes of your research talk:

1. *Begin with an interesting image or clip.*
   A provocative image that relates to your research is an interesting way to begin. Alternatively, you could display a 1 min movie clip that demonstrates a physical phenomenon or explains the relevance of what will follow. Remember that the first slide is often displayed before the talk begins. That means the audience will be looking at that slide for the longest amount of time. Thus, your first slide is an opportunity to be compelling, illuminating, or to inspire curiosity.

2. *Begin with a provocative statement.*
   This beginning is easy if you have a very important discovery to convey which is provocative in itself. In this case, do not beat around the bush, just say what you have done. "The reason Western cultures conquered the Americas is because they had better guns, germs, and steel." "We've discovered the earliest members of the human race." "We invented a new technique for turning off individual neurons in the brain (optogenetics)." In most cases your discovery may be more modest. Nevertheless, there may be aspects of your research that go against the audience's expectation. For example, your research findings may challenge current orthodoxy, such as Emily Oster's talk on HIV in Africa (Oster, 2007) "Everything we know about the spread of HIV in Africa is wrong." In these cases, provocative statements can be very effective.

3. *Tell a short personal story.*
   Anecdotes are one of the oldest forms of engaging an audience. Anecdotes or personal stories help make an idea memorable by engaging the listener in something personal. Margot Leitman (2015) suggested that beginning with a personal story will win the audience to your side. Using an anecdote as your beginning can be powerful, but you must use it carefully. Theresa MacPhail (2015) had four suggestions when incorporating an anecdote: (1) Don't make it all about you. (2) Weave the anecdote throughout the

presentation. (3) Find specific details with which your audience can identify. (4) Reference the story again at the end of the talk. Without question, your anecdote should connect to your research. If you can find the right personal story to tell, you will create the opportunity for a memorable moment.

4. *Create an analogy.*
Analogies can provide a nice frame for your central research idea. Unlike a standard, straightforward explanation of concepts, analogies provide poetic connections between commonly understood ideas and the complex information you want to convey. A good analogy is a type of extended metaphor and can help you gracefully and efficiently introduce challenging concepts (Monarth, 2009).

5. *Use an interesting prop.*
Is your research about something concrete or tangible? Can you bring in a sample? If you are studying the effect of a disease on apples, you could bring examples of healthy and diseased apples for the audience to see. Allowing your audience to see or touch the subject of your research will be more effective than describing that object theoretically.

6. *Create a moment of audience participation.*
Audiences enjoy participating in the construction of knowledge (Conner, 2013). Can you find a way to include the audience in the setup of your research talk? Perhaps you could ask for a volunteer and use them to demonstrate a point. Or ask the audience to look around and observe something specific in the room. Engaging the audience from the first moment in your presentation shifts their cognition from passive to active, and if executed well, could create some critical initial buy-in. A note of caution about polling the audience: if you ask an easy or obvious question, your introduction could fall flat. It is more effective to ask a thought-provoking question and allow your audience the time to contemplate the answer.

7. *Begin with an astonishing statistic.*
"Every year, three times as much garbage is dumped in the oceans as the weight of fish caught." Do you happen to have an unusual or impressive statistic that would be a strong introduction to your research talk? Like analogies, statistics can provide a bridge to the relevance of your research idea. This is especially true if you can relate the statistic to the audience's experience or give it a meaningful context. For example, a 12% increase in crop yield may not mean much to someone who is not a farmer. Is 12% a lot or a little? Relating this crop yield to the total income of the farm, indicating whether it would make or break a growing season, or comparing this result to those from other more well-known products can help.

8. *Set up a mystery that you will solve.*

   If you are researching the solution to something specific, a mystery might be a nice frame for your talk. For example, "Rita has been struggling with infertility. She also suffers from celiac disease. Is there a link between damaged villi and reproductive fertility? My lab is working to find out." This scenario sets up an ideal opportunity for a story, and automatically creates emotional connection to the subject.

9. *Make it personal.*

   What is it about your research that is important to you? How did you determine your particular research focus? For some, an area of research is connected to someone specific or a particular life event. If you feel comfortable and can do so genuinely, share this information with your audience and let them know how much you care about your research subject. "This is my grandmother, Ada. She's 72 years old. She suffers from Alzheimer's. Currently we don't understand the specific genetic mechanisms that cause Alzheimer's. My research is working toward…"

10. *Make them laugh.*

    Are you a natural comic? Do you enjoy making your friends laugh? Nothing will charm an audience like beginning your talk with a funny moment. Humor is enjoyable. Laughter is contagious. But take care. If you are not good at jokes, if you fail to deliver punch lines properly, this may not be the right choice for the beginning of your talk. If you are interested in incorporating humor, remember that telling jokes is only one form of comedy. Puns, sarcasm, running gags, or tongue-in-cheek observations can also tickle an audience. The most important thing to consider when incorporating humor is that the style of comedy matches your personality, and the injection of humor is natural, not forced. Using humor is not for everyone, but if you can, we highly recommend incorporating some lighter moments.

Here is how Itai applies these concepts to his research talk:

Itai begins his talk on flight of the fruit fly with this introduction: "This is Drosophila melanogaster. Isn't he beautiful?" (#1, #2, #10). He then proceeds to demonstrate a typical controls problem by balancing a meter stick on the tips of his fingers—balancing an inverted pendulum (#5). In longer talks, he uses volunteers to do this demonstration (#6). Finally, he makes the analogy that, similar to an inverted pendulum flapping flight is unstable, so that fruit flies need to make constant corrections to their wing strokes to maintain their stability (#4). However, because the time scale for these corrections are related to the size of the object, flies must perform these corrections on the millisecond time scale (#7), how do they do this? (#8) He then finishes by telling the audience that this talk is about understanding this control reflex.

This beginning demonstrates ways in which you can incorporate multiple strategies to create a good beginning. Within about a minute, Itai has hooked the audience, and they want to know more about his research.

Exercise 4.1 Creating a Good Beginning

1. Using your Core Message and your Dramatic Arc as reference points, brainstorm about some possible beginnings for your talk. Try to fill in at least one example in each of the following categories. Do not worry if some of them seem ridiculous or improbable. Just freely brain storm and see which possibilities emerge:

   - An interesting image or clip
   - A provocative statement
   - A short personal story
   - An analogy
   - An interesting prop
   - A moment of audience participation
   - An astonishing statistic
   - A mystery you will solve
   - A personal reason for this research topic
   - A moment of humor

2. Narrow down your choices to your strongest possibilities. Map out how you could use each of these strategies to begin your talk. Write out some dialogue if you find that helpful. Get on your feet and practice them in front of a camera. Once you have had a chance to review your practice sessions, polish the strongest version. Your beginning should be the most rehearsed component of your research talk (Fig. 4.1).

Fig. 4.1 Constructing a great beginning

Strong Endings

Equally important to a good beginning, and often overlooked, is the importance of a strong ending. The purpose of ending strongly is to create a sense of emotional satisfaction for the audience, make the talk memorable (in a positive way), and fill the audience with a sense of satisfaction akin to just having finished a great meal (Meyers & Nix, 2011). A strong ending reminds your audience of what they learned, and why it is important. Ending well helps the audience understand why they should care about your research talk. A weak ending, on the other hand, can deflate energy and momentum, dissipate your connection with the audience, and make you feel less positive about your performance. Therefore, it is important to take the time to figure out how you want to end your presentation.

In order to achieve the final emotional payoff for your research talk, you need to *wrap-up* the ideas you have just delivered, and *signal* to the audience that you have finished your presentation. The wrap-up helps the audience remember the important points you just talked about, while the signal helps the audience determine when the talk is over.

The Wrap-Up

In your dramatic arc, the wrap-up is the resolution section of your story. Here is where you fill in any narrative gaps in your storytelling and provide concluding information about something you have addressed earlier in your talk. Once again, we encourage you to try something different from the traditional approach while still staying true to the message you are trying to convey. Typical endings consist of a conclusion slide with a few bullet points summarizing your main findings. This slide is often followed by one listing the collaborators and grants. While this format is useful for communicating your information efficiently and is often a safe fallback, it is not very engaging. Below are a number of suggestions for improving your wrap-up (Fig. 4.2).

1. *The circle back or callback.*
   A callback is a reference to whatever you used to hook your audience at the beginning of your talk. If you used an anecdote, find a way to talk about it again. If you included a joke, perhaps you can remind the audience of that funny moment. If you began with an image, you could wrap up with the same image on the screen.

Fig. 4.2 Strong endings

2. *Make them think.*

   Talk about the implications for your research. Most speakers assume that the audience will understand why their research is important. After all, to you it is obvious. But remember that the audience has only been thinking about this topic for the duration of your talk. During your talk most of the audience's effort is consumed with trying to keep up as you speak. It is unlikely that they will have enough time to connect the dots themselves. Be clear about why your research matters, and where it fits in the big picture.

3. *End with a memorable image.*

   Images make a stronger impression than a collection of words. If you can find an image or video that summarizes the core message of your talk, the audience will have better recall. Some phenomena are naturally beautiful. In that case, simply putting up an image of the phenomena will remind the audience of the beauty in a breaking drop of water, or the intricacy of a metamaterial's deformation, or the vibrancy in a work of art. Choosing an appropriate image is important. If you put up an image of a cat, they will remember a cat. It will be memorable, but at the expense of your message. Instead, use simple and effective images that summarize your findings and core message.

4. *End with a cliff-hanger.*
   Sometimes, your talk is a progress report and will not have a natural end-ing. Your research may be ongoing, or you may have finished one of several related projects that contribute to a bigger body of work. You can still entice the audience by giving them a road map for the next steps in your research. The important thing is to provide them with the vision for your future work. Many students are hesitant to adopt this approach because they are afraid they will be held accountable if these future projects fail. Audiences, however, are often willing to be flexible with their expectations, especially during a talk's conclusion. Everyone realizes that until the paper is written and published, everything in a talk is subject to some amount of revision. So you have some leeway. In fact, you can use this description of your future plans to solicit ideas from the audience, or even collaborations that could facilitate your future work.

A traditional conference talk on the Flight of the Fruit Fly would end with a slide containing the following summary bullet points:

- Just as fish use lift forces when swimming, flies also use drag forces when flying, so flying and swimming are not so different.
- A fly's ability to correct for midair perturbations can be modeled using PI controllers.
- Using genetic manipulation, we are determining the neural circuit that takes sensory signals and converts them to muscle contraction and wing actuation.
- Evolution has pushed the development of halteres making Diptera more maneuverable, better predators, and better able to avoid predation.

This information is accurate and provides a nice summary of primary points, but it is dull and far from memorable. Instead, Itai uses some of the strategies above to make the end of this talk more interesting and compelling.

1. *Callback.* (31:10)
   Itai uses the image of the fly and returns to the sentence with which he began his talk "This is Drosophila melanogaster, isn't he beautiful?"
2. *Make them think.* (30:40)
   Itai shows an image of a fly on the patent for the Wright Brothers' flier. This image is meant to recall for the audience the elegant strategies insects use to control their flight. He then asks the audience to imagine the world 350 million years ago before insects took flight. "There were no trees, there were no flowers, all plants were less than three meters tall. The ability of insects to take to the air and control their flight had a profound effect on our planet's ecology."

3. *End with a memorable image.* (31:10)

   Itai begins and ends his talk with a photograph of *Drosophila melanogaster*. The beauty of this image is that it is initially revolting. By the end of the talk, however, the audience falls in love with this creature and the image evokes a completely different response. By using the same image, the audience appreciates where they started and where they ended up.

4. *Determine the best place to end the talk.* (31:10)

   To achieve a strong ending Itai thanks his collaborators *after* he delivers his ending. Once the applause has begun to die down, he immediately adds his acknowledgment slide and says his thank-yous. This approach creates an emotionally satisfying ending to the research talk while still finding a way to thank contributors. Another option for preserving a punchy ending is to thank collaborators at the beginning or middle of the talk.

   This wrap-up demonstrates ways in which you can create a strong conclusion to your research talk. This approach does not require any more time than the traditional conclusion slides and leaves the audience with a greater sense of emotional satisfaction.

Exercise 4.2 The Wrap-Up

1. Think about why you are doing this research. What are the ultimate implications? What do you find so cool about what you have discovered? Why are you putting in so much time to solve this puzzle? These are the questions that should guide the construction of your wrap-up. If you have already constructed a traditional concluding slide, you can use the bullet points as guidelines for what you want the audience to take away from your talk. Then brainstorm about some alternative ways of presenting the same ideas. Using the prompts below, experiment with possible wrap-ups to your research talk.

   - Callback: Go back to the first few slides and determine if there is a theme you can return to in your ending.
   - Make them think: What are the implications of your work? If what you say holds up, how will it change the world? We recommend that you consider the largest possible implications and then scale back from there until you feel comfortable with the connection you are making.
   - End with a memorable image: There are many images that could work here. Try using an image from your first slide. Such an image would also address the callback. Create a compelling image that summarizes your findings. This is the last impression you will leave on your audience so choose the image carefully.

- End with a cliff-hanger: This strategy works well if the talk you just gave is part of a larger story with an even bigger impact. If this is the case, letting the audience glimpse what is to come can be very effective. Think about the next steps in your research project. How will the pieces for this larger story come together? What are the implications if your future research leads the project one way or the other?

2. Once again, narrow your choices to your strongest possibilities. Experiment with the different endings and see which ones are most effective at keeping your audience riveted to the very end of your talk.

The Signal

At the end of every talk, good speakers will signal the audience that their story is coming to an end. There are some techniques that can be used to accomplish this task:

1. Take a short pause before the final sentence.
2. Slow down while delivering the final sentence.
3. Lift the pitch during the first half of the final sentence, and then drop it during the last half.
4. End the final sentence with a recognizable ending phrase like "Thank you."
5. Accompany your last words with a gesture, such as dropping your hands to your side or a slight nod of thanks to the audience. (See more on the use of the body in Chap. 8)

In Itai's talk, he has three closing moments that help signal applause. (31:09) First, he pauses, and makes direct eye contact with the audience. Second, he repeats the introduction "This is Drosophila melanogaster, isn't he beautiful?" Notice the pitch drop at the end of the sentence. Rather than making this a question as he did at the beginning of his talk, with the pitch drop he turns it into a statement signaling finality. Finally, he takes a short pause and ends with a firm but pleasant "Thank you."

Exercise 4.3 The Signal

1. Identify the final sentence of your research talk. Practice pausing before delivering this sentence, lifting the pitch at the beginning and dropping the pitch at the end.
2. Choose a final tag out phrase like "Thank you." "That concludes my remarks." "I'll take your questions now." "I hope you've enjoyed this journey as much as I have."
3. Practice delivering your wrap-up and concluding with your signal sentence and tag out phrase. Record your delivery and perfect it until you can conclude your talk easily and smoothly.

Beginnings and endings are critically important for your research talk. The audience will use the first few moments of your presentation to decide if listening to what you have to say is worth their time and whether or not you are credible in your delivery. You will know if you have a great beginning if the audience sticks around for the rest of your talk. Strong endings, sometimes overlooked and thrown away, are as important as good beginnings. The way you end your talk determines the impression the audience will take away with them. You will know when you have nailed the ending if the audience stays for the applause rather than rushing to catch their next talk. These two parts of your talk are where you have the greatest creative license to experiment. They will be anchored by the rigorous content in the body of the talk. So be brave. The audience will reward you for it.

5

Make Them Care

Having worked hard to streamline your talk and create an engaging beginning and a strong ending, it is now time to ask yourself once again: Why should anyone care about this research? Does your talk answer this question? If not, more effort may be required to explain why your work is relevant to your audience members. To some scientists and scholars, selling their work is considered dirty and belittling. After all, if the work is great why does one need to sell it? It should sell itself. It should be obvious why it is important. The answers to these questions require us to overcome the "curse of knowledge" (Heath & Heath, 2007). It may be easy for *you* to connect the dots and recognize the implications of your research on a grander scale, but a typical audience member will need some help. Certainly there is a risk of overselling your work. Underselling your work by not connecting the dots for your audience is also a problem. You may need to experiment to get the balance of your message just right.

It may be helpful to remember that the goal in giving a good talk is to hook the audience. In many ways your presentation is an advertisement for your work. For many listeners, it will serve as an entry point to a field or topic they are interested in understanding. If they like the talk, they will be more likely to read your papers. If they like the papers, they may want to collaborate with you on a project or enter the field themselves. Therefore, rather than trying to cover every single aspect of your work, your primary objective is to make your audience care about this specific result.

This goal can be accomplished by covering fewer topics with greater clarity. Greater clarity means that your audience will be able to discern whether your result is substantial. It is better to be clear and honest about your findings

© Springer Nature Switzerland AG 2019
I. Cohen, M. Dreyer-Lude, *Finding Your Research Voice*,
https://doi.org/10.1007/978-3-030-31520-7_5

rather than hiding them through obfuscation. In the best-case scenario, you have a rigorous well-developed ground-breaking result and by clarifying your presentation you are making it easy for people to understand the importance of your research.

One of the key strategies for connecting with your listeners is to appeal to their self-interest. Jerry Weissman (2008) called this WIIFY—What's in it for you? An audience becomes actively engaged when they can see how what you are talking about matters to them personally or professionally (Heath & Heath, 2007). The strategies you will use to create audience buy-in may depend on the particular problem you are studying. If you are studying a disease or tracking the value in a social change movement, making your case may be easy. Who would not want their corn to be more nutritious or to hear how the social climate is improving? In the case of malnutrition and medical diseases in particular, while including personal anecdotes and concrete examples can be powerful, it is important to make sure you are not exploiting those who are suffering and that your descriptions are authentic. Some experimentation may be in order to get your script just right. If your work has the potential to usher in a new type of important technology or asks us to reconsider the implications of a moment in history, you might try inspiring the audience by describing an imagined future or a reimagined past. In such cases, a good rule of thumb is that you should be as optimistic as possible without misleading your audience. In most cases, your research will address one component of a broader strategy. While it is important to describe the big-picture goal, it is also important to provide context for where your particular piece of the puzzle fits. Making the connection between what you are researching and its ultimate impact on society is a good way to make your audience care.

The following exercise demonstrates how to think in new ways in order to determine why your research may be important to your audience.

Exercise 5.1 Make Me Buy This

1. Identify two or three volunteers who will demonstrate the exercise for the rest of the group.
2. Assign the roles of salesperson, customer, and coach. (If you only have two people use salesperson and customer.)
3. Have the audience choose an unusual product for the salesperson to sell. It is important that this item is either an odd or challenging product. For example, a roll of toilet paper, one shoe, or your half-eaten sandwich.
4. Have the audience choose a role for the customer: unemployed father of three, farmer, or captain of a spaceship.

5. The salesperson now has two minutes to try and sell this product to the customer using a variety of tactics.
6. The coach (who can also serve as time keeper) may provide suggestions when the salesperson gets stuck.
7. During the 2 min time frame, if the salesperson convinces the customer of the value of the product, the customer indicates a "win" by shaking the hand of the salesperson and saying "I'll take it." If a tactic does not achieve a win and the customer is not buying, the salesperson must shift tactics and try again and again. To achieve success, try to imagine how your product will serve the specific needs of this customer. Try not to get stuck on obvious uses for your product.
8. Once the demonstration is complete, split into teams of two or three and repeat the exercise on your own.
9. Keeping the same teams, repeat the exercise using your own research as the product. The person whose research is being "sold" gets to choose the role for the customer (research advisor, grant manager, reporter, or tax payer). The other group member(s) should try to help by providing ideas for how to make your research relevant to your chosen audience (Fig. 5.1).

Fig. 5.1 Make me buy this

Another strategy for making your audience care is to emphasize the aesthetic elements in your work. You could, for example, relate the motions of the electrons you are studying to a beautiful dance. You could demonstrate the impact of an economic initiative through colorful graphics. You could present images of the landscapes that inspired Whitman's work as you analyze the contemporary relevance of his writing. Perhaps the phenomena you are investigating is visually striking. You can find good examples of arresting images in the gallery of fluid motion, or by using other sources of stock imagery like iStock or Getty Images. Another strategy is to illustrate a clever or elegant experimental study aimed at elucidating a complicated puzzle. You might highlight a particularly ingenious experimental apparatus you have designed, or a beautiful mathematical framework you have developed for investigating a particular phenomenon. You could provide striking images of artifacts that are relevant to your anthropological study. These latter strategies often work well if you are giving a talk to specialists who already care about the topic and know why it is important. Many of your audience members will have gotten into their field because they thought it was beautiful. By appealing to their aesthetic sense, you can remind them of why they became scholars in the first place.

If you are still stumped, a good starting point is to determine why *you* are passionate about your research. Something drives you to get up each morning and wrestle with your research problem. Was it a nagging sense that our current picture of the world is not quite right? Was it about hunting down an idea that kept tugging at you? Or perhaps a quixotic quest to elucidate some behavior that enthralls you? Communicating that personal passion during your talk is another strong way to establish a connection. Your passion will inspire their passion. Research has demonstrated that passion is contagious (Gallo, 2014). If it appears that you are indifferent to your topic, it is unlikely that you will generate interest from your listeners (Leitman, 2015). Passion, on the other hand, creates presence, and presence is interesting to watch. The following exercise provides some suggestions for finding an emotional hook for your talk:

Exercise 5.2 Finding the Emotional Connection

1. **WIIFY**. How is your research specifically important to your audience? Why will it matter to them? Is it personally or professionally important to those who will be listening? Do not assume that they will know this already.
2. **Aesthetic appeal**. What is it that is beautiful about your work? Did you discover an elegant way to solve a problem? Perhaps there are images that are just beautiful. Maybe there is an elegance that you can highlight. Remember that the aesthetics may relate to the streamlined explanation rather than the tortuous path you took to get to your results.
3. **Personal passion**. Take a moment to reflect on what is personally important to you about your research. Is it how your findings will help others? Is it the intellectual puzzle you will solve? It is the joy of discovery? Late in the process

of conducting research, after months or years of scholarly work on your subject, it can be challenging to remember what sparked your interest in the first place. Write down some thoughts about why you are personally passionate about your work. Then weave this passion into your delivery (Fig. 5.2).

Fig. 5.2 Working on the emotional connection

In his research talk, Itai uses several tactics to help the audience understand why they should care. First, he helps them develop an emotional relationship to the fruit fly. He does this by speaking about the object of his research with passion, affection, and admiration. By the end of his talk, his audience is sad to say goodbye to this underestimated creature. At several points during the talk Itai will pose a rhetorical question as though he has anticipated the thoughts of his audience. He then proceeds to answer this question by directly addressing the issue. Throughout his talk, Itai draws parallels to real-world applications—particularly to flight and the engineering required to make it possible. Some members in his audience will have been on an airplane multiple times in their lifetime. Because this is a science audience, Itai makes clear the impact his research will have on future developments. His work has real-world applications and lays the groundwork for future studies in this field.

If you want to make an impact with your research presentation, you need to connect with your audience and make them feel something. Scholarly presentations are often about data, arguments, and concepts. This reality can be a challenge to engaging your audience. Long-term memory is constructed by emotional experience. Memory athletes use this idea to generate emotionally charged stories that enable them to recall long card sequences (Dresler et al., 2017). Within the context of a conference where the brain is exposed to lots of ideas, facts, and figures from many different presentations, the audience will have a better chance of retaining ideas that are emotionally compelling. Stimulating retention by making your audience care will significantly enhance the impact of your talk.

Part II

The Performance

6

Calibrating Your Speaking Presence

The audience has come to hear *you* speak, the real you, not a phony constructed version of you (Howard & Tivnan, 2003). To learn to be authentic and poised as a public performer, you do not need to construct another persona, but you may need to make some adjustments to your physical performance. It is important to recognize that some of what you do as a performer is already working for you and that some of it is not. In the same way that it is hard for us to hear our own verbal ticks until they are pointed out to us, the first step towards reconnecting with your physical expression is to learn to recognize the signals you are sending. Because many of us have been raised in an atmosphere that is critical and judgmental—think of all those oral exams and graduate school presentations—we have learned to defend ourselves by reducing our vulnerability and restricting our expression. Unfortunately, these behaviors often get in the way of fully communicating your story and conveying your own personality.

It may be useful to remind yourself of the broad range of emotions that you can access when trying to make a point or tell a story. The following exercise is an exaggerated version of the broad palette that is available to each of us.

Exercise 6.1 Tell It Like...

1. Have each participant write down a mundane piece of news that they will share with the class. This news should be about three sentences long. It could be a weather report, the latest on the economy, or road closings. It is helpful to the exercise if the content is rather dull.

© Springer Nature Switzerland AG 2019
I. Cohen, M. Dreyer-Lude, *Finding Your Research Voice*,
https://doi.org/10.1007/978-3-030-31520-7_6

2. On several small slips of paper, have each participant write down an emotionally charged circumstance through which someone could deliver this news. For example, you just won the lottery, your goldfish just died, your hateful stepparent is coming for a visit, you found out that you need to have a tooth pulled, you have been asked to bungee jump by your best friend... Here it is helpful to choose circumstances that generate extreme emotional states.
3. Put the slips in a hat or other container. Have everyone blindly choose one of the slips of paper.
4. One at a time, have participants stand in the front of the room and deliver their "news" to the class using the circumstances indicated on the chosen slip of paper. Each performer should be as clear as possible with their tone of voice and body language.
5. Have the group guess what the circumstances might be based on the performer's behavior.
6. Have the performer share with the group the circumstances they were trying to convey. Ask the group for feedback regarding other strategies the performer could use to make those circumstances clear. Could the performer use more body language, facial expression, or a different tone of voice?

One of the things that makes this exercise humorous is the mismatch between the mundane content of the "news" and the way in which this information is delivered. When the signals you are sending do not match your intended message, the audience is often left confused and uncertain about how to respond. If you are jumping with excitement when you are describing something unremarkable, the audience may think you are crazy. On the other hand, if you talk about your main result using the same cadence and volume as the rest of the presentation, the audience will not understand that this part of the talk is of particular importance. Detailed feedback about nuance may require a professional. But you should be able to understand the basics on your own or at least with some feedback from a colleague. With these ideas in mind, let us look at your 10 min talk and determine if there are glaring inconsistencies between the content you are presenting, your intended message, and your delivery.

Exercise 6.2 Finding Your Disconnect

1. Gather in groups of three or four.
2. Play the clip of one group member's 10 min talk.
3. As a group try to identify where you see a possible disconnect between the research message and the speaker's physical and vocal expression.
4. Review each member's 10 min talk in the same way.
5. Do you notice any common problems? What was working well in each talk that you can borrow for yours?

Another useful exercise is to learn to recognize how others use physical expression in their talks to strengthen their message. Once you can recognize these behaviors, you can learn to mimic them and incorporate them into your own performance. Through borrowing and testing these elements in your own talks, keeping the ones that work for you and throwing the rest away, you will develop your own style (Fig. 6.1).

Exercise 6.3 Borrowing and Stealing

Go through Itai's talk, attend a departmental seminar, or watch a talk of your choosing on the internet and identify physical expressions or other behaviors that you find particularly effective for communicating an idea. A good test would be to ask yourself whether that behavior helped you understand a key point of the talk. Identify one or two elements that you can try out in your own talk. Determine where in your talk you could insert this element and then try it out on a test audience.

Recognizing how you are expressing something is different from being able to change it. The above exercises are examples of the ways in which you can explore, test, and adjust your range of expression. This process may feel foreign at first, but effectively incorporating these elements can make a big dif-

Fig. 6.1 Analyzing your behavior

ference in how the audience receives your talk. The next two chapters will address in greater detail the two most important aspects of your delivery: using your voice to be heard and understood, and using your body to support your ideas through posture and physical expression. These chapters and the accompanying exercises will help you tailor and broaden your tool set for self-expression. We encourage you to experiment beyond what you would naturally do to determine what style works best for you.

7

Using Your Voice

The way you speak controls how your presentation will be received. If your audience cannot hear you, finds your voice quality unpleasant, cannot understand your words, cannot keep pace with your delivery, or in some other way becomes frustrated by the way you speak, you will have failed to engage, connect, and inspire. It is helpful to recognize the qualities of good speaking and to identify what is working and what is not in your speaking voice. When speaking well in front of an audience, you are *audible* (using adequate *volume*), *comprehensible* (we can understand your words and what they mean because you are using good *diction*), and *sonorous* (using appropriate *pitch* and a pleasant *tone*). In addition, good speakers use an appropriate *pace* and remove distracting *fillers*.

Volume

Being heard when you speak is critically important. If your audience cannot hear you, they will not pay attention. It can be exhausting to strain to hear a speaker's words. Volume is also a critical indicator of confidence and energy (Stone & Bachner, 1994). If you are speaking softly, it appears as though you do not believe in what you are saying or in yourself as the speaker (Reynolds, 2012). On the other hand, it is important to refrain from shouting (Reynolds, 2012). Shouting can make you sound angry. The goal is to speak loud enough to be heard with a nice sense of energy and enthusiasm. Learning to calibrate your volume is an important component of good vocal expression.

© Springer Nature Switzerland AG 2019
I. Cohen, M. Dreyer-Lude, *Finding Your Research Voice*,
https://doi.org/10.1007/978-3-030-31520-7_7

Calibrating your own voice can be tricky. What you hear inside of your head is different from what others hear when you speak. When you make a volume adjustment, it might feel as though you are shouting or whispering, but this is probably not true. Your inner perception of volume may not align with the actual sound you are making in the room. A recording of your presentation or feedback from peers will help you calibrate the volume of your voice.

It is often surprising to discover how loud one actually needs to be in order to be heard. For some, speaking this loudly is not natural and can create vocal strain. In order to avoid damaging your voice when trying to speak more loudly, it is important to review some of the basics of voice production. Volume is created through breath not musculature. You cannot make your voice louder by pushing the sound out through your throat (Stone & Bachner, 1994). Pushing or striving for a louder sound creates tension that may result in laryngitis or vocal nodes. Rather, think of the process of creating more volume as releasing the tension, taking a deeper breath, and sending out a bigger idea. Voice specialists Patsy Rodenburg (2015) and Kristen Linklater (1976) recommend a version of this cycle of events: (1) Release physical tension. (2) Let the breath drop down. (3) Think of what you want to say. (4) Send the energy out. Rodenburg describes the importance of breath support:

This support of air should connect to and with the voice. Support starts and stops sound as we increase or cut off the air supply, just as you experience with an air hose used to inflate one of your car tyres [sic]. Think of the sound of the voice as starting with the support from the centre [sic] of the body and *not* the throat (2015, p. 152).

Exercise 7.1 Learning to Control Your Volume Using Your Breath

1. Place one hand on your belly and one on your chest. Notice where you are breathing. Ideal breath support comes from belly breathing. Many people breathe primarily from the chest. This limits the amount of air you draw into your lungs and the amount of breath you have to increase the volume of your voice.
2. To help you learn to draw your breath from your belly, try the following: Push all of the air out of your lungs—every drop. Wait one moment until you feel the need to breathe. Give in to the need to breathe by allowing your belly to draw the breath in. Do this several times so that your body begins to remember to draw the breath down.
3. Now place your hand on your belly. Imagine that your torso is cavernous and hollow but elastic. Allow the breath in naturally (do not push), but imagine it coming all the way down to your pelvis and filling that space. Then allow the breath to leave your body on its own. Practice this breathing method several times.

4. Keeping your hand on your belly, look at someone in the room and tell them "Hey!" Imagine that the "Hey" comes from your belly (not your throat muscles).
5. Yawn and feel your throat open.
6. Visualizing that your throat is relaxed and open (as in the yawn), find another person, let the breath drop down, and then tell them "Hey!"
7. To practice volume control, choose a partner. Stand two feet away from one another and say "Hey!" at a volume that is appropriate. Now move ten feet apart and adjust your volume using your breath. Now move 20 feet apart and say "Hey!" Think of these volume adjustments as using energy (supported by your breath) to throw the words at the other person—do not push them. Rather, release them like a ball when you toss it.

Once you begin to understand how to generate volume with your breath, it is important to adjust the level of volume you use when speaking in public. The following exercise helps you learn to calibrate your volume to meet the needs of the room.

Exercise 7.2 Volume Calibration

1. Try to find an empty lecture hall or space comparable to a conference room.
2. Have everyone stand at the back of the hall.
3. One by one, have each participant walk to the center of the speaking stage and introduce themselves, as in "Hello. My name is xxx and I study xxx."
4. If the group at the back agrees that this person was audible, they may return to the back of the room. If not, the speaker must continue to repeat this phrase, increasing the vocal energy until the other members of the group agree that the volume is appropriate.
5. Make sure everyone takes a turn to have their volume production checked.

Microphones are a big help with volume control when you are speaking to a crowd. Learning to use a microphone effectively is an important contribution to the impression of competence and confidence. Microphones can also make it easier to use lower volume for dramatic effect without losing comprehension, as you can move in closer to a microphone when you drop your volume and still be heard. When using a microphone, it will still be important to use good breathing technique to project your voice, though the volume calibration will be different.

You will probably be using one of two types of microphones: a fixed mic on a lectern, or a lavalier mic that is attached to your body. In both cases, do a mic check before you speak to be sure the mic is working. All microphones have a range within which they can pick up your voice. Think of this as an invisible cone that extends from the microphone. The better the microphone, the wider the cone. With a fixed microphone, you want to position the mic so that you can stand up straight and be heard whether you are looking at your notes, or the audience. Having your voice go in and out of microphone transmission can be distracting, and can undermine the quality of your presentation. Take the time to adjust the mic before you begin.

The other kind of mic, a lavalier or lapel mic, simplifies voice transmission and allows you to move more freely as you present. Lavalier mics are normally attached to your clothing in the front and connected to a power pack that is clipped to your belt or waistband. If you know you will be using a lavalier mic in your presentation, it is helpful to wear clothing that will accommodate these devices. With a lavalier mic you must remain conscious of the transmission cone as well. Place the mic far enough away from your face to allow for a larger transmission area, preferably in the middle of your chest area. If the mic is too close to your neck, your voice could jump in and out as you turn your head. To avoid these pitfalls, the best strategy is to practice using this technology.

Exercise 7.3 Using a Microphone

1. To determine the range of a podium microphone, begin to speak and move your head left and right to determine when you move in and out of amplification. Now speak again and increase and decrease your distance from the microphone to determine the ideal relationship between your mouth and the mic.
2. To determine the appropriate position of a lavalier microphone, attach the microphone to different positions on your shirt or dress to determine how the amplification changes. Turn your head from one side to the other and notice any volume changes. Remember, the ideal position would be in the mid-chest region for most microphones.
3. To avoid feedback, try to stay away from any speakers.
4. Try going through some of your talking points and determine whether you are staying within the microphone's range.

In the best-case scenario, there will be someone on hand to help you adjust the microphone position and resulting volume. Practicing with a microphone beforehand will allow you to make such adjustments on your own even when a technician is not available.

Diction

If you speak loudly enough to be heard, you might still struggle to be understood. Diction refers to the clarity with which we speak. Some speakers mumble, some speak with an accent, or some only encounter clarity issues on certain words or sounds. The easiest way to improve your diction is to practice pronouncing consonants effectively. Be careful not to turn "t"s into "d"s, and be sure to enunciate your "r"s. In English, vowels provide the substance of sound but consonants divide the sound into packets of meaning. The good news is that comprehensible pronunciation of English words is a skill you can learn. There are a number of books containing explanations and exercises that can be helpful. Below is an exercise to help you determine where you may need help with diction. Remember that if you are a non-native speaker and

---

Exercise 7.4 Diction

1. A tight tongue makes it difficult to pronounce things clearly. Here is an easy exercise you can use to loosen up your tongue. Imagine that you have just put a large spoonful of peanut butter in your mouth (any imaginary sticky substance will do). It is now stuck in every crevice and you must use your tongue to clear it out. Get under, behind, between every part of your mouth using your tongue, pretending to clean out the peanut butter. This will naturally encourage you to stretch your tongue in every direction.
2. Now that your tongue is loose, say the following tongue twisters out loud to determine which sounds are a struggle for you. Pay attention as you read the phrases out loud. If you struggle to clearly articulate a particular kind of sound, this will alert you to a deficiency in your pronunciation. A partner in this exercise can help you focus on the sounds that can be improved for clarity.

Plosives: (p, b, d, t, k, g)
Picky people pick Peter Pan Peanut-Butter, 'tis the peanut-butter picky people pick.
A big black bug bit a big black bear, made the big black bear bleed blood.
Begging beguilingly.
Begrudging curmudgeon.
Curiously obscure procurer.

Fricatives: (v, f, th, s, z, sh)
Alluvial bivalve.
Four furious friends fought for the phone.
Very well, very well, very well …
Three free things set three things free.
Scissors sizzle, thistles sizzle.
Casual clothes are provisional for leisurely trips across Asia.
Three tethered teething things.

Affricate: (ch)
Chester Cheetah chews a chunk of cheap cheddar cheese.
Imagine an imaginary menagerie manager.
Charming bachelor Chuck.
Charting challenging channels.

Nasals: (m, n, ng)
Aluminum linoleum.
Belonging longer.
Mary Mac's mother's making Mary Mac marry me.
Nine nice night nurses nursing nicely.
The king would sing, about a ring that would go ding.
How many mahogany and mohair hassocks has Hermione?

Frictionless: (r, l, y, w)
Ray Rag ran across a rough road. Across a rough road Ray Rag ran. Where is the rough road Ray Rag ran across?
Luke Luck likes lakes.
Yellow yams yank Yolanda's yak.
I wish I were what I was when I wished I were what I am.
A wooden worm wouldn't be worthy of worship but would he if he wondered and worried about what he would be worthy of if he wasn't wooden?
Red leather, yellow leather.

    Once you've determined which elements need work, you can find exercises (in books or on the web) that will help you build proficiency with those particular sounds. When you develop the muscles that produce these sounds, you will find them much easier to pronounce.

3. Perform the first ten minutes of your talk to find the sounds that you need to work on in your presentations. Again, enlist the help of a partner to identify particular words or sounds that lack clarity. Are there particular words that are difficult for you to enunciate? Does your partner look confused when you say a particular word? Get feedback from your partner on how to pronounce a word more clearly. Practice the specific tongue twisters listed above to help you generate the sounds needed to pronounce these words.

your words are already comprehensible, there is no need to remove your accent. In fact, some audiences appreciate the new melody and tones a foreign accent can provide.

Pitch, Tone, and Prosody

Pitch, tone, and prosody control the emotional qualities of speech. They also contribute to comprehension and the appearance of confidence and authority. Pitch refers to the range of expression of your speaking voice and could be correlated to a musical scale (between middle C and F, for example). Tone refers to the quality of the sound. A round, resonant tone is more pleasant for the listener than a harsh, abrasive tone. Prosody refers to the melody the speaker uses to convey the meaning of the statement. The use of these ideas when speaking is often cultural. Tonal languages like Mandarin place the same word on various pitches using different melodies to connote distinct meanings. English has no grammatical pitch requirements, although English speakers also use pitch and prosody to convey meaning. In English, pitch is used to distinguish the difference between a question and a statement. We raise the pitch at the end of a sentence if we are asking a question and lower it if we are making a statement.

Operative words are important in many languages and provide a useful communication tool. The operative word is the word (or words) in a sentence that are most important. In English, the speaker employs operative words using a combination of pitch and emphasis. The sentence, "The apples in this basket are red" can have slightly different meanings when the speaker shifts emphasis to the operative word. "The *apples* in this basket are red" (not the oranges) versus "The apples in *this* basket are red" (not the other basket) versus "The apples in this basket are *red*" (not green). The words are the same. The meaning shifts with pitch and emphasis. Understanding the power of operative words may come in handy in your research talk. By carefully selecting particular words at important moments, you can help your audience understand where and when to pay attention.

A verbal tick associated with pitch that can confuse the listener is Upspeak. Upspeak occurs when you turn a statement into a question by raising the pitch at the end of the sentence. Upspeak is a habit that should be eliminated as soon as possible, as it suggests uncertainty and lack of confidence (Stone & Bachner, 1994). If you hear the sentence, "I'll meet you after work tomorrow." as "I'll meet you after work tomorrow?" you might not be certain that the person speaking will actually show up. When persuading an audience with your research talk, you want to convey confidence, certainty, and authority. Upspeak can undermine those qualities and leave the audience with the impression that you are not sure about your findings.

It is also worth considering your optimal pitch placement. Pitch placement is the note you use most often when speaking. Everyone has an optimal pitch that is determined by their physiology. Gender can influence pitch placement. Some women tend to speak above their optimal pitch (because they may believe it to be more feminine); some men speak below their optimal pitch (because they may believe it to be more masculine). These are often unconscious decisions that become embedded habits or habitual pitch. A lower than optimal pitch reduces audibility because the sounds tend to blur and distort. A higher than optimal pitch can become shrill or breathy and undermines the perception of confidence and competence. Here are two exercises that can help with optimal pitch and Upspeak.

Exercise 7.5 Pitch

1. <u>Optimal pitch</u>. Working with a partner, speak a sentence in what you would consider to be your normal pitch.
2. Repeat the sentence using a higher pitch.
3. Repeat the sentence using a lower pitch.
4. Discuss with your partner which version sounds the best (to him or her) and feels the most comfortable (to you). Does the ideal version match the pitch you are currently using? Or do you need to make an adjustment?
5. <u>Upspeak</u>. Speak the sentence you used above as though it were a question, by using a higher tone for the last word in the sentence. Now repeat the sentence as though it were a statement by dropping the tone at the end. Do you hear the difference? If not, try having your partner repeat what they heard, emphasizing the tonal difference at the end of the sentence.
6. On your own, record yourself describing some of your slides. Notice whether you are using any Upspeak. If you find that you do, take a paragraph of text and read it through while converting every sentence into a question (this sensitizes you to the habit). Then read it through again and make every sentence a statement (even if there are questions in the text). Note if speaking in statements is easy or difficult for you. Begin to pay attention to the ways in which you might be using Upspeak in daily conversation. Ask friends to point it out to you when it happens. Removing a habit requires awareness and time.

A rich, resonant tone is often a component of those voices we admire most. A beautiful tone is open, full of sound, and easy to hear. Consider the speaking voices of Helen Mirren, Michelle Yeoh, Mahershala Ali, Salma Hayek, or Jon Hamm. There are a handful of problems that can impede a rich tone. The first is vocal tension which can make your voice sound tired or strained. Vocal fatigue occurs when you are engaging your vocal chords unnecessarily as you speak. If you find that you often lose your voice after public speaking engagements, you may need to learn to relax your throat muscles and let the breath do the work. One version of vocal tension manifests in something known as vocal fry. Vocal fry is the process of dropping your pitch to its lowest register, tightening your chords, and speaking with a scratchy sound (listen to Bradley Cooper, Amy Adams, and Mary-Louise Parker as examples of vocal fry). Vocal fry is hard on your voice. If you have fallen into the habit of using vocal fry and are finding that you regularly lose your voice due to vocal tension, removing vocal fry from your speaking pattern may help.

Calibrating your breath is also important for generating good tone. Some speakers put too little breath into their vocalization, which creates vocal tension. This tension can occur when the throat is too closed, the chest is tight, and you are not using belly breathing to support your voice. When your throat is closed, the muscles in your throat must work extra hard to generate volume, resulting in a strained tone. To alleviate this situation, you need to open your throat and allow the breath to flow freely. Try yawning to open your throat and be sure to use your belly breathing to control your volume, not your throat muscles. At the other end of the spectrum, some speakers do not engage their vocal chords enough, allowing too much air to pass through without generating sound. This can create a soft, breathy tone (think Marilyn Monroe). A breathy tone that saps the volume from your voice can suggest a lack of confidence or authority. If you tend to be breathy when you speak, you need to learn to connect your breath more strongly to your voice to generate a fuller sound. Breathiness can sometimes arise when a speaker is nervous. Should this occur, it is important to take a moment to breathe deeply and relax, letting your breathing slow down a bit (perhaps by taking a sip of water). This will allow you to return to generating a nice rich tone with your speaking voice. Of course, using a breathy voice or a harsh tone might be the *right* choice for making a certain point in your presentation and can always be inserted for dramatic effect.

The following exercises are most useful when done in pairs. For each of these exercises, engage in the activity by taking turns as reader and responder. The responder can help the reader recognize when she or he is employing the habit or leaving the habit behind.

Exercise 7.6 Tone

1. Vocal Fry. Find a paragraph of text to read. Read a portion of it trying to deliberately use vocal fry. Now read it trying to relax your vocal chords and leave your throat open and released.
2. Breath Control. Read the paragraph tightening your throat and releasing as little breath as possible when you speak. Follow that by reading the paragraph in a breathy tone à la Marilyn Monroe. Give yourself a moment to shake off these extreme choices and read the paragraph with a balanced tone.
3. Apply these ideas to your presentation. Choose a particular moment in your talk and try speaking with a balanced tone. If you are ambitious, try to determine if you can generate other effects by experimenting with your tone.

Pace

Speaking at an appropriate speed can be critically important to comprehension. It is rare for presenters to speak too slowly. It is far more common to hear them speak too quickly. The information you are conveying is familiar to you, but it is brand new to your audience. Each sentence you say will have new information your listeners need to process. If you speak too quickly, the audience will spend some of their processing time returning to the previous sentence to try and decode what they think you might have said. It can be challenging to learn to slow down when you normally speak quickly. When you do manage to slow down, it may feel as though you are speaking at a glacial pace, but this is often not the case. Your version of slow may be just the right speed for your audience.

In addition to tempo, it is helpful to pay attention to rhythm when you are speaking. Rhythm is comprised of the *pattern* of sound you make when you speak. Speeding up slightly when you are delivering information of less importance and slowing down a bit when you reach a critical moment in your talk can add variety and aural interest. Think of the rhythm of your talk as grouping words together into blocks of meaning (Meyers & Nix, 2011). A predictable rhythm can lull the audience into a semiconscious state, while an unpredictable or changing rhythm can keep them on the edge of their seat wondering what you will say next.

One of the best tricks to use when shaping the rhythm of your talk is the dramatic pause. A dramatic pause halts the action and allows a moment of silence before the talk continues. There are multiple places to include a pause. (1) Right after you enter and before you begin to speak. (2) Directly before presenting an important finding. (3) As you transition from one section of your talk to another. A pause, particularly before a transition, can give your audience a moment to reflect and consider what you have just said, and to begin to build links between what you have said and what you are about to say (Arnold, 2010). Well-placed pauses can help reinforce your dramatic arc, building tension and interest as you move through your points one by one.

Exercise 7.7 Pace

1. <u>Slowing down</u>. Find a paragraph of text and record yourself reading it at your normal pace. Now record yourself reading the paragraph at what you would consider a very slow pace. You could try thinking of yourself as reading the text under water, or in a slow-motion movie. Play the "normal" and the "slow" recordings for a colleague and ask for their opinion of the pace or

tempo with which you are speaking. Which do they prefer? Which recording is more comprehensible? Are there further adjustments that they recommend for the pace of certain words or phrases?

2. Add variety. Find a section of text that includes two or three paragraphs. Read through it once to get a sense of the flow. Take two or three colored highlighters and code the text, indicating which sentences you will read more quickly because they are not crucial (perhaps coded yellow), which you will emphasize by reading more slowly (perhaps orange), and those that you will deliver at a normal pace (perhaps blue). Once you have coded the text, read through it, trying to follow the guidelines you have created. On the first pass through the text, exaggerate your choices so that fast and slow are significantly different from one another. On the second pass, soften your choices and try to make the transitions sound more natural and conversational.

3. Adding pauses. Using the same text, read through the document and flag at least three places where you could put a pause. You will pause either to highlight a point, to reveal an important moment, or to help you transition from one section to another. Record yourself reading through the text again, keeping the rhythmic variety you coded in the previous exercise, and adding the pauses you have now identified.

4. Compare recordings. Find a colleague and play the original (part 1) and final (part 3) recordings. Ask for feedback on the effectiveness of each recording.

5. Apply ideas to your research presentation. Go back to your 10 min talk and try to identify places where you can implement these ideas to achieve the desired effect. We often suggest that you focus on three places in your talk where you can consciously remember to change the pace to generate dramatic effect.

Fillers

Many people include fillers in their speech. Some common examples are: like, um, man, so, right, ok, totally, well, you know, you see, I mean, I guess… Any word or collection of words can become a filler. We use fillers to cover insecurities when we are speaking, to buy time when we are thinking, to pad our sentence and attempt to add substance, or to try and create a smooth subject transition. Fillers are usually unconscious habits that can be very distracting for your audience (Kushner, 2010). There are a variety of reasons why fillers may have crept into your vocabulary. We recommend taking the time to remove them from your speech patterns. The best way to remove a filler from your research talk (or your everyday conversation) is to enlist some colleagues to alert you when you have just used a filler while speaking. By pointing out the use of a filler word the moment you say it, your colleagues are helping you make an unconscious habit conscious. Habits are hard to break unless you can catch yourself in the act. Once you become conscious of these vocal habits, you can begin to remove them from your vocabulary. Stone and Bachner (1994) recommended saying nothing when you have the impulse to add a filler and to allow the silence to serve as an opportunity to think of the next thing you have to say.

Exercise 7.8 Fillers

Review an earlier recording of your talk. Use a sheet of paper to record a hash mark each time you use a filler. When you have finished watching and listening to your talk and making your marks, look at your sheet of paper and notice how many fillers you used. Which specific fillers are problematic for you? When do you tend to use them? Identify patterns in your use of fillers so that you can begin to break those habits.

"A strong pleasant voice is the greatest asset a speaker can have" (Stone & Bachner, 1994, p. 19). In this chapter, we have provided an introduction to the basic components of a good speaking voice. Most importantly, you must be heard and understood. Calibrating your volume, diction, and pace will get you most of the way there. As you become a more proficient speaker, you will find that playing with your pitch, tone, and dramatic pauses will provide nuance and specificity that will allow you to engage with your audience in a more meaningful way. If you attend to each of these elements of your voice, you can develop an exceptional instrument that will be a pleasure to hear.

8

Using Your Body

The goal in mastering body language is to create a sense of presence—a quality that compels the audience to look at you in a positive way. Presence creates the expectation that something extraordinary is about to happen. According to Sarnoff and Moore (1987), presence is generated through the ways in which we carry ourselves. They have identified a number of factors that can affect the perception of whether or not you have presence: your physical bearing, the way you enter the room, the quality of your support material, the atmosphere in the room, and the makeup of the audience. Only some of these factors are under your control, and body language is one of them.

Studies have made clear that we are communicating with our body language and facial expressions whether we realize it or not (Ekman, 2007; Key, 1975). Folded arms can telegraph emotional distance, a slouched posture suggests a lack of confidence, and failure to make eye contact can make you look deceptive (Burgoon & Saine, 1978; Scheflen, 1972). To convey confidence and poise when presenting, you will need to become aware of your body language. Ideal physical behavior is *relaxed* (free of tension and full of ease), *confident* (upright and certain), and *expressive* (energized and passionate). The goal, however, is not to force these behaviors. Relaxed, confident, and expressive body language should emerge naturally from your preparation and practice. Once prepared, you will feel more relaxed. If your story is well formed, you will tell it in a more confident manner. Your emotional connection to your material can naturally lead to a passionate and expressive performance. This chapter will focus on how to develop these qualities in your presentation style and help you strengthen your message through the use of your body.

© Springer Nature Switzerland AG 2019
I. Cohen, M. Dreyer-Lude, *Finding Your Research Voice*,
https://doi.org/10.1007/978-3-030-31520-7_8

In our experience, many of the issues associated with miscommunication through body language arise from bad habits: rocking when you stand; fidgeting with your pencil; looking at the floor when you speak. Even overuse of a "good" gesture can be a problem and become distracting. On the other end of the spectrum, standing perfectly still without conveying any physical expression is also a problematic habit. Some of these behaviors are easily fixed. Recording yourself and watching your behavior will often alert you to the most egregious problems. Other bad habits may be more ingrained and will take longer to correct. The first step to fixing these more entrenched behaviors is to become aware of them and consciously correct them so that, over time, the good behaviors will become the unconscious habit.

Appropriate body language during a presentation can be broken down into several components: posture, gesture, eye contact, facial expression, and movement. Mastering each of these aspects will help make you more confident about your presentation. Physical behavior—including posture and gesture—can change how we feel emotionally (Gallo, 2014, p. 94). A slouched, withdrawn posture can contribute to feelings of sadness and lethargy, while an open, upright posture can help generate feelings of strength and contentment. A strong, confident posture and hand gestures also create a positive feedback loop with your audience. If you appear more confident, the audience will believe you are confident, helping you feel more confident, creating a virtuous cycle.

Posture

Posture is a key component in creating presence. A few examples of different body postures can be seen in Fig. 8.1. In the first image (sway back), the model has pushed his pelvis too far forward. This posture disturbs the natural center of gravity and puts a strain on the lower back and thighs. In the second image (lumbar lordosis), the pelvis is kipped in the opposite direction. Although the spine has a natural curve toward the tail bone, exaggerating this curve creates excessive tension in the lower back muscles and results in lower back pain. The third image (thoracic kyphosis) shows a collapse of the upper spine and is common for those who work hunched over a desk for many hours a day. This posture deflates the overall physical energy and puts strain on the shoulder and neck muscles. The next image (forward head) has most of the elements of a good posture, but the forward head position creates tension in the neck muscles, which could result in vocal strain and the loss of your voice. In the final image in the illustra-

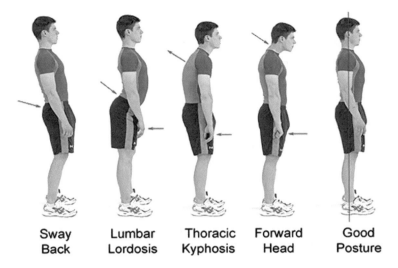

| Sway Back | Lumbar Lordosis | Thoracic Kyphosis | Forward Head | Good Posture |

Fig. 8.1 Posture

tion above (good posture), you will notice that the body lines up along a vertical plumb line. The head is lifted but not thrust forward, the chest is lifted but the shoulders are released, the posture is tall but relaxed and free of tension. These ideas are the foundation of a number of movement methodologies (Maisel, 1989; Feldenkrais, 2013; Jakubowicz, 2018). The key to good posture is the release of tension and the proper realignment of the body. Proper alignment will allow your body to function more effectively, and learning to release tension is key to creating a relaxed stage presence. Here are the components of a relaxed, tension free posture:

1. You are standing with your feet shoulder-width apart and your weight evenly balanced.
2. Your knees are soft, not locked.
3. Your pelvis is rooted, but free.
4. Your chest is lifted while your shoulders and arms are free of tension.
5. Your head is floating up and free. The back of your neck is released and long.

It may take time to master a perfect posture, particularly if bad posture is your habit. Try to incorporate a lifted, relaxed posture into your everyday routine. Before long it will become your new habit.

Exercise 8.1 Posture

If possible, do these exercises in front of a mirror, or in collaboration with a colleague who can provide a reflection for you, either through a cell phone or through verbal feedback.

1. Standing posture.
    (a) **Feet**: Stand with your feet shoulder width apart. Your feet may point forward or at a slight angle, whatever is natural for you.
    (b) **Balance**: Check your center of balance by rocking in pendulum fashion left and right and then front and back, allowing the movement to become smaller until you can sense the central point of weight balance.
    (c) **Knees**: Check to be sure that your knees are soft and not locked. Soft knees are not bent knees, and their slight bend should be mostly imperceptible to the observer. Locked knees can cut off circulation and make you feel faint. Keep your knees released but not bent.
    (d) **Pelvis**: Take a look at the position of your pelvis. Rock your pelvis very far forward and then very far backward. Feel the tension that each of these extreme positions create, particularly in your lower back. Now think of rooting your tail bone by tilting it toward the ground so that it releases your lower back. The tilt should not be so extreme that it pushes your body out of alignment. If you are doing this properly, you should feel the tension in your lower back soften and disappear.
    (e) **Shoulders and chest**: Look at the position of your shoulders and chest. It is common for the upper chest to be collapsed forward a bit and for the shoulders to sag forward as well. This position compromises your appearance of confidence and erodes your physical energy. To fix this posture problem, imagine that you have a string tied to the center of your chest that extends upward at a 45° angle. Now allow that string to pull your chest up gently. Your shoulders should not do the work by pushing backward, they should just follow along and remain released and relaxed. Lifting your chest in this way should feel gentle and easy, not strained. You may find that this position helps with your breathing as well.
    (f) **Head**: Take a look at how your head is positioned on the top of your body. Are you pushing it back or jutting it forward? To find the proper position, imagine that your head is light as a balloon and would float to the ceiling if it were not attached. Allow your head to float up. There is a natural tendency for your chin to float up as well, creating tension in your neck. Let the movement of your head initiate from the back of your head so that your chin will naturally float down (Maisel, 1989).
    (g) **Neck**: Put your hand softly on the back of your neck. You want to release any residual tension that remains here. Gently direct your neck muscles to soften into your hand. You should feel a slight release and softening of these muscles.

(h) **Release**: Move your head around gently continuing to imagine that your head is as light as air. Check to be sure that you have not added more tension. Use your hand to release any residual tension.

(i) Review the video of your initial talk. Identify whether you are using good posture and if not, make adjustments.

2. Sitting posture.
   You can practice an upright, relaxed posture even when you are sitting.

   (a) Sit in a chair with both feet on the floor.
   (b) Allow your position to be well back in the chair but not all the way to the back.
   (c) Place your pelvis in a rooted position. Do not allow it to collapse.
   (d) Lift your chest as instructed above. Make sure that your arms and shoulders are released.
   (e) Check your head and neck and allow your head to float up and forward as instructed above.
       (This posture will allow you to sit tension free for hours, although regular breaks are best when working at a desk.)
   (f) If you wish to rise from your chair without adding tension, simply tip your torso forward and send the energy up through your head at a 45° angle. Allow your body to follow that direction and you will easily rise from your chair.

3. Releasing tension.
   It is difficult to convey confidence if you are feeling tense and rigid. Use the following exercise to help you relax before you speak.

   (a) Sit in a chair (or you may stand) and place your body into proper posture.
   (b) With your eyes open or closed (depending on what circumstances will allow), start at your head or at your feet, and work your way through your body, focusing on one area and gently instructing it to release. For example, direct your attention to your feet, notice where there might be tension, direct your feet to release that tension and to soften into the floor (and then continue working your way up your body).
   (c) Once you have released the tension in your body, stand as instructed above.
   (d) Check to be sure you are in an upright, tension-free posture and make corrections if necessary.
   (e) Now begin to walk around, trying to keep the tension released from your body. You may begin at a slow tempo, but try to work your way up to a normal walking pace, still walking without tension.
   (f) There may be parts of your talk where you feel extra tension (e.g., beginning of the talk or during the question period). Try implementing the tension-relaxation tips listed above. Releasing tension in your posture may help you relax a bit and think more clearly (Fig. 8.2).

Fig. 8.2 Working on posture

Gesture

Another key component of physical self-confidence is the use of gesture when you speak. Goldin-Meadow (2005) offered a technical definition of gesture as those movements of the hands that directly relate to what you are saying, but gesture lends itself to a broader definition. Gesture may be considered any component of your physical expression that you employ to highlight, emphasize, or clarify what you have to say. We will focus primarily on hand gestures, but you can also gesture with your head, your feet, your posture, your hips, or any other part of the body that is appropriate for expression. What if you do not feel comfortable using your hands to express yourself? Do you really need to incorporate gesture? Gallo (2014) says yes, because gestures help clarify ideas and contribute to an impression of confidence. Using supporting gestures when speaking in public adds credibility to what you are saying.

It is important to be sure that your gestures match your words. If your gestures do not appear to support your text, the audience will become uncomfortable and might begin to distrust your message (Gallo, 2014). For those who struggle with gesture, the real question is how to do it. First and foremost, allow gestures to emerge spontaneously from your own natural movement. The exercise below will help you become more comfortable with using gesture. The best placement for gesture in a public speaking environment is within the Gesture Box. The Gesture Box refers to the area in front of you that is located above the waistline and below the shoulders. This is considered the power area (Gallo, 2014). When the hands reach too low (below the hipline), the speaker appears awkward or stiff. When the hands reach too high (above the shoulders), the speaker appears overly enthusiastic. You can break the Gesture Box rule when you want to make a particularly important point, surprise the audience, or change the rhythm, but in general the Gesture Box is the area to use when gesturing.

For those who wonder how and when they might incorporate gesture into their talk, Arnold (2010) offered some useful suggestions. These are not prescriptive, but they might help you generate ideas for when to use gesture:

1. You can DESCRIBE something using your hands.
2. You can ENUMERATE with your fingers.
3. You can EMPHASIZE with a specific gesture.
4. You can CHOP instead of pointing.
5. You can REACH OUT toward the audience.
6. You can MODEL and show the audience what they should do in your example. (Arnold, 2010, pp. 49–50) (Fig. 8.3)

We recommend the following rules for incorporating gesture in your talk:

1. **Keep your gestures free flowing** but not overly energized. Gestures need to support your statement but not draw too much attention to themselves.
2. **Resist the overuse of any gesture** as it will lose its effectiveness in supporting your message. An overly repetitive gesture shifts from providing support to being a tick that can become irritating to watch (Collins, 2011).
3. **Watch the hands in pockets**. Hiding your hands in your pockets can betray nervousness and/or look overly casual. One hand now and then is not a problem, but be sure not to fiddle with keys, coins, or anything else you have stored there.
4. **Do not use pointers, pens, or chalk when gesturing**. Put them down first. Lasers can be hard on the eyes, so be careful not to point them at the

Fig. 8.3 Working on gesture

audience. Also, watch for excessive laser circling of something in your visual material. Point to it once, and then trust that the audience is with you.

5. **Watch for nervous ticks**. This includes but is not limited to: fidgeting, tapping, jingling, or standing rigidly in place (Gallo, 2014).

Exercise 8.2 Gesture

1. Find a short story from your recent past that you can tell your colleagues. Think through the details from beginning to end.
2. Now tell the story to a small group without using words. Try to use only your hands to convey your narrative. Note: This may be difficult and you may find yourself resorting to absurd measures to communicate. That is ok. Do not censor yourself. The goal here is to become comfortable communicating with your hands.
3. Now tell the story again, and add words to the story. Although your need for absurd gestures should significantly diminish, continue to include gesture to illuminate and support your points. Your colleagues should press you to continue using your hands to explain your tale, even when only words will do.

4. Shift to a new small group and tell your tale. Use as much or as little gesture as feels appropriate. Allow the use of gesture to feel natural, as though these movements belong to you.
5. Finally, look through your talk for moments where gestures could support your communication. Practice incorporating gesture into those sections of your research talk.

Eye Contact

Looking another person directly in the eyes is an act of courage. If you are having a conversation with someone, and they refuse to look you in the eye, in US academic culture one might assume that they are (1) lying to you, (2) nervous about something, or (3) socially maladjusted. A key element of confidence, whether you are speaking with one person or to a crowd, is the courage to look your audience in the eye. You can help establish an emotional connection with your audience through your gaze. Eye contact helps your audience feel important, which will help them set their other activities aside and focus on you (Zeoli, 2008). Eye contact helps create an atmosphere of intimacy between you and your audience. Eye contact also conveys power and confidence (Stone and Bachner, 1994). If you have the courage to look your audience in the eye, you must really believe in what you are saying.

There are a number of recommended "Dos and Don'ts" when making eye contact with your audience. One of the most important things to do is to make sure that you spend most of your talk looking at your audience rather than the projection screen or your notes (Reynolds, 2012). Keep your eyes up and out. Experts recommend that you keep your eyes on the audience 90% of the time (Sarnoff and Moore, 1987). If you are managing to keep your eyes up during your talk, you need to watch that you do not fall into the habit of sweeping or scanning the audience (Meyers and Nix, 2011). Connect with one audience member at a time, delivering a point here, and a point there, as you shift your gaze from one audience member to another. Try to shift your gaze evenhandedly. Create a pattern that will help you include everyone: center, right, left, center, right, left. Be sure to choose new people when you shift your gaze. Try not to cling to those three people who seem to be enthusiastic about your presentation. Look at all areas of the audience and find someone with whom to connect. If looking an audience member directly in the eye overwhelms you, consider talking to their nose or their forehead. Be sure, however, not to focus your gaze far above their heads (Kushner, 2010). They will not be fooled and you will fail to make the connection. If it is dark in the room and you are illuminated, it is still important to look out to all areas of the audience. It can be challenging to make eye contact as you deliver your research talk, but remember that it is critically important for several reasons: conveying confidence, establishing an emotional connection, and helping your audience feel as though what you are saying matters.

Exercise 8.3 Eye Contact

1. To warm up your ability to make eye contact, gather in an open area of the room.
2. Approach each person in your group one at a time and shake their hand. As you do, look them directly in the eye and say "Hello." Be sure to greet everyone in the room.
3. Now choose a five-minute section of your research talk, or a short story you can share with your peers.
4. One at a time, stand and deliver your text to the group making an effort to individually connect with each person in the room.
5. Offer one another feedback when you are serving as an audience member. Did the speaker genuinely connect with you? Was the gaze long enough? Did the speaker include everyone or fall into a pattern?
6. The next time you deliver a talk in public, have a colleague watch your presentation and offer feedback on your use of eye contact. Keep this feedback in mind for your next talk and adjust as necessary.
7. A note for international conference presentations: the use and interpretation of eye contact might differ in other cultures. Be sure to check if possible with a native from that land (Fig. 8.4).

Fig. 8.4 Using eye contact to connect

Facial Expression

An expressive face helps your audience understand what you mean. We know from the research on mirror neurons that your audience will not only be inferring your point of view when they watch you speak, they will also, at least to some degree, be living it with you (Niedenthal, 2007). You can encourage the audience to enjoy your presentation by enjoying it yourself. Allow your facial expression to reflect pleasure and you will encourage the audience to find pleasure in what you say.

You are not expected to deliver a Tony Award-winning performance with your facial expression. The most important expression you can incorporate into your presentation is a smile. Even in the most serious of professional situations, a light smile will help you convey confidence. A smile creates rapport in almost every situation (Kushner, 2010). It can be challenging to smile when you are nervous or concerned about how your talk will be received. Smiling may help you feel happier about what you are doing. Research has shown that if you come across as warm and friendly, you will appear to your audience as more competent (Fiske, 2018). Practice smiling and it will come more naturally to you. If you already smile regularly, do not hesitate to include it in your presentation style.

Exercise 8.4 Facial Expression: The Smile

1. If you are someone who normally appears somber, serious, or even stern, your facial muscles will be accustomed to this expression as a resting pose. In order to create a more expressive range, you will need to "teach" your face another possibility. Try to take a few moments during your day to pay attention to your facial expression and incorporate a smile. You do not need to flash a big, toothy grin, just use a light smile as your target. If you are unsure whether or not you are smiling, check in the mirror and adjust the degree of smile accordingly. It is important that the expression you use be recognizable as a smile to outside observers. This "smile practice session" can be as long or as short as you like. We recommend repeating it until it feels more natural. Initially, this exercise might require attention and concentration, but eventually, a light smile will come easily to you.
2. Review a recording of your talk and determine whether you smile. If you have gone through the full talk without smiling, try to identify moments in your talk where a smile would be appropriate. Then try incorporating a smile during those moments the next time you give the presentation (Fig. 8.5).

Fig. 8.5 The smile

Movement

In addition to posture, gesture, eye contact, and facial expression, learning to move appropriately can contribute to the appearance of confidence. Proximity, orientation, and how to move are all aspects of this subject.

Edward Hall (1963) introduced the idea of proxemics in the 1960s. Proximity is the distance between you and your audience. Whether or not you have control over this distance will depend on the space in which you are presenting. Ideally, you want to remove obstacles between you and your viewers to create more intimacy and enhance your emotional connection. One of the key issues in many locations is whether or not you will be tethered to a podium. If the room is large and you require a microphone, and the only microphone available to you is hardwired to the podium, you may be forced to remain there. However, if you have a lavalier mic, or the room happens to be small, come out from behind the podium.

It is also important to orient yourself so that you are facing your audience whenever possible. It is a mistake to spend much of your presentation talking to the projection screen. You can turn to the screen to point out a specific element on a particular slide, but afterwards be sure to return your body's orientation to the front. You are the star of this show, not your slide presentation.

Incorporating a stage cross during your presentation can help you look confident and in charge (Reynolds, 2012). You could use a stage cross to help emphasize an important point by drawing attention to that moment. Depending on your presentation space, you might consider choreographing these moves into your talk. Once you have determined where you want to move, it is important to motivate your cross. Here are some strategies for incorporating a motivated cross into your talk.

1. Cross to the other side of the stage to reference a point on your projection screen. This will seem justifiable to your audience if you are identifying something on that side of your image.
2. Cross in order to talk to a specific person or section of your audience. To do this, as you begin to transition into this section of your story, make eye contact with someone in that section, and let your need to talk to him or her draw you over.
3. You can cross back to your notes (perhaps on the podium?) in order to quickly reference something (even if you do not need a reminder). This will place you in a neutral position once again from which you can choose to move to another location on the stage.
4. You may also consider crossing during the question and answer period of your presentation by approaching the side of the stage closest to your questioner.
5. Be sure to find moments of stillness as well. Crossing the stage should be the periodic accent to your talk, not the dominant element.

There are a few aspects of movement that you want to avoid. Be careful not to rock or bounce when you are standing. These distractions are nervous habits and should be removed from your presentation. When you cross, do not float purposelessly from one place to another (Meyers and Nix, 2011). Rather, find your target location, create the reason to go there (your motivation), and cross strongly to that position. When you are standing, stand on two feet in a relaxed posture. Avoid standing on one leg, or crossing your feet. Keep your posture upright and relaxed with an even balance of weight, and you will be perfectly positioned to make your next stage cross.

Exercise 8.5 Adding Movement

1. Take a look at the shape (the dramatic arc) of your research talk. Identify specific moments when you can cross (a) to the projection screen, (b) to the audience, or (c) back to the podium or desk.
2. Practice including this movement by following the instructions above. Allow your peers to provide feedback regarding how natural and appropriate your movements choices were.

Entrances and Exits: How to Take the Stage

"Taking the stage" in the theater is the process of stepping in front of the audience and commanding attention (Howey, 2005). Even in a conference presentation you are participating in a theatrical ritual. People have gathered; someone will come to the front and tell a story; those in the audience will listen. The audience wants you to radiate confidence and charisma as you prepare to speak. That is why they have come. You can signal from the very start of your presentation that you are prepared, confident, competent, and excited to share what you have to say. We spoke earlier about the importance of beginnings when you construct a story, as the beginning is where the audience decides whether or not your story is worth their time. What you may not know is that the audience makes some decisions about you even before your story begins. The evaluation of you as a speaker starts with your introduction, not with the first moment of your story. So it is important to get this pre-beginning section right.

The first step in making sure your entrance goes smoothly is to check your equipment earlier in the day. It is important to manage the logistics of your environment, as unnecessary fumbling or malfunctioning equipment can undermine the impression of confidence (Collins, 2011). This short investment of time can make a big difference between a strong first impression and the sense that you are not sure of what you are doing. If you know that your equipment is set up and ready to go, this will remove at least one worry from your presentation and set you up for a successful beginning.

Often (but not always) you will be introduced to the audience before you speak. This is an important moment in your presentation. The audience will be scrutinizing you during your introduction, making small judgements about your appearance and your confidence level. This may sound intimidating, but it is inevitable. The good news is that you can control many aspects of the impression you make so it is important to begin conveying confidence from the very first moment. While you wait to move to the front of the room, stand or sit tall and relaxed, find a comfortable position for your hands, place a light smile on your face, and keep your eyes fixed on the speaker.

Once you have been introduced, you need to take the stage. This is another important moment in your presentation. You are providing information to the audience about what to expect, through your body language. The way you approach the front of the room makes a difference. Again, theatrical tradition provides some insight. "When [actors] enter the room, the energy level rises. You perk up, stop what you're doing, and focus on them. You expect something interesting to happen" (Halpern & Lubar, 2004, p. 1). You are the actor in this scenario, and the audience will be watching to see if you convey presence. Stage presence is communicated through a combination of factors. First, be sure that you are tall and relaxed (this is where the posture lesson comes into play). Next, focus your gaze forward toward where you are going, not on the floor. Look either at the person who introduced you or the place where you will be speaking. This will motivate your movement and create the impression that you know exactly where you are going. As you walk, keep your facial expression pleasant. Remember that a light smile conveys confidence. The tempo with which you move is also important. Your pace should be brisk and purposeful, energized but unhurried (Dale Carnegie Training, 2011; Stone & Bachner, 1994). A brisk pace is one that is slightly faster than a normal walking pace, but not so fast that you appear to be rushing. A simple approach to the front of the room is actually a complex combination of messages that you convey to your audience, all of which are under your control.

The moment you arrive on the stage and before you begin your talk is crucial. Many public speaking experts agree that to convey confidence and relaxation, once you reach the stage, *do not immediately begin to speak*. You have a few options before you begin. If appropriate, shake the hand of the person who introduced you (conveying warmth and openness), and thank that person for the nice introduction (grace and charm). Then move to the place where you will begin your talk and make any final adjustments to your equipment. Once that is finished, take a breath, look up, and make eye contact with your audience. All of this may take only 15 s, but those 15 s are important. By taking your time, you create the impression that you have done this before, that you have the situation under control, and that the audience is in good hands for the next short period of time. The very first moment of your talk once you are settled is a *moment of silence*. A silent moment is a confident moment. A silent moment allows your audience to relax and get ready to listen. A silent moment begins the process of anticipation and helps build tension. Do not immediately speak when you reach the stage. Take a moment and allow the audience to settle their attention on you.

You should pay similar attention to your exit. After you have delivered your compelling research story and have answered audience questions, you need to exit gracefully. In the section on endings, we discussed how to signal the end of your story and to encourage applause. As your beginning has a pre-beginning (your entrance), your ending has a post-ending (your exit). Think through how you will exit before you begin your presentation, perhaps during the time you spend checking your equipment. What moves will you need to make? Do you need to take a computer or notes with you? Do you need to return the lavalier? Do you need to log out of a software program in preparation for the next speaker? Ideally, you want to gather and return things as quickly and smoothly as possible, and then confidently leave the stage. The movement rules of your exit are the same as your entrance. Move briskly and confidently using a pace slightly faster than normal but not hurried. Eyes are forward and up looking toward your goal (the back of the room? the exit door? a place where you will sit and listen to other speakers?). You should have a light smile on your face whether or not you feel you did well. Leave the stage with grace and pride and that is what your audience will remember about you.

Exercise 8.6 Entrances and Exits

1. One by one practice entering and *preparing* to say the first line of your story. Have another member of the group introduce you so that you can practice this part of your pre-beginning. As you work, have the group give feedback on the following:

    (a) Was your demeanor relaxed and confident as you were being introduced? Did you keep your eyes on the speaker?
    (b) Was your approach to the podium brisk but unhurried? Were your eyes up and forward? Did you smile as you entered?
    (c) Once you were on stage, did you take a moment to thank your host, check your equipment, take a breath, and make eye contact with the audience?
    (d) Finish the exercise by saying the first line of your presentation.

2. Now practice your exit. Begin with the final sentence of your story and then move into your exit. Again, have the group check for the following.

    (a) Did you smoothly negotiate the logistics of exiting (returning the mic, disconnecting your laptop, etc.)?
    (b) Was your walk brisk but unhurried? Were your eyes up and forward and focused on where you were headed?
    (c) Did you smile as you left the stage?

There are many aspects of a confident presentation, and the way you use your body plays an important role. You can control some aspects of how people view you. You can shape their perceptions of your competence, and the message you choose to send about who you are as a speaker. Learning to control your body language and the impression you make in public will help you charm your audience, engage them in your subject, and persuade them that what you have to say is worth their while.

Part III

The Event

9

Dealing with Stage Fright

We are often asked about how to eliminate nervousness before delivering a talk. Being nervous is common, even for seasoned performers. What professional speakers have learned is that nerves are a part of the job. A lack of nervousness can create a sloppy or half-hearted presentation. That is because nervousness is a form of energy, energy that you can harness and funnel into your presentation. Nerves can be your ally. Nerves can sharpen your focus and heighten your attention. Rather than resisting your feelings of nervousness, consider reframing how you think about the butterflies you feel before you speak. Think of nerves as that extra punch that will add passion to your talk. Shift your worry about your upcoming talk to a focus on the value of having all of that extra energy. Speakers who appear to have more charisma on stage have probably learned to channel their nerves into passion for their topic (Howard & Tivnan, 2003).

It is also helpful to remember that nervousness often dissipates during your talk. Once you have begun your well-prepared presentation, you can forget about the frightening part of public speaking and focus on the message you want to tell (Dale Carnegie Training, 2011). That is why professional speakers spend the greatest part of their preparation time working on the first few lines of their presentation. They know that starting a presentation well allows the speaker to shift more quickly from a focus on nervousness to the telling of the story.

In reframing your relationship to nervousness, it is also helpful to reframe your relationship to your audience. Those who came to hear you speak did so because they were interested in what you have to say. They want you to shine, to be compelling, to change their world view with your new discoveries

© Springer Nature Switzerland AG 2019
I. Cohen, M. Dreyer-Lude, *Finding Your Research Voice*,
https://doi.org/10.1007/978-3-030-31520-7_9

(Kushner, 2010). Audience members are usually on your side. Keep that in mind as you step onto the stage to share your research.

One of the biggest reasons people become nervous about speaking in public is the fear of making a mistake. It is the rare public presentation that does not contain at least one mistake. Mistakes happen. You might bungle your words, skip ahead in your slide presentation, forget to mention something critical, or trip as you enter. There are many grim possibilities and spending time imagining them or worrying about them is time and energy wasted. The audience is not usually that bothered by little mistakes. The only person who really cares is you (Zeoli, 2008). Your audience can and will forgive any mistakes you make, as long you are willing to let them go and move on (Stone & Bachner, 1994). If you make a mistake, get yourself back on track as quickly as possible and continue with your presentation. If you forget it, they will forget it. They are not interested in focusing on small imperfections. What interests them is hearing your story.

The number one antidote to extreme anxiety when giving a research talk is *preparation*. It is the rare person who can "wing it" when speaking about something important in public. Preparing a sensational research talk takes time. You must start weeks ahead constructing and perfecting your story. Once you have created an interesting story to tell, you must begin practicing your presentation. You cannot do well if you have not adequately rehearsed your talk (Meyers & Nix, 2011). We highly recommend recording your later practice sessions so that you can see and assess your performance from an objective point of view. Once you have polished your talk, find opportunities to present it to a practice audience and invite them to offer you feedback. If possible, try to practice in an environment similar to the location where you will be presenting. The closer you can come to recreating the actual presentation circumstances when you practice, the more comfortable you will feel (Menzel & Carrell, 1994). Regardless of the tools you use to improve your performance, ample time for practice is key. It can be challenging to find the time to practice. Every minute you prepare will be well worth the effort. Not only will the final product be more convincing, it will be more fun to present your research, and you will naturally feel more confident while delivering your message.

Mentally preparing for your presentation is another helpful antidote to anxiety. There are a variety of techniques you can use to get your mind settled and ready to go (Fig. 9.1). Belly breathing and the tension/relaxation method can help you settle before your presentation. Belly breathing is the process of sitting in a chair and allowing your breathing to slow and deepen so that you are breathing more from your belly than from your chest. The tension/relaxation method involves

Fig. 9.1 Learning to relax before a presentation

isolating specific muscle groups, deliberately tensing them, holding that tension for a moment, and then letting the tension go as you completely relax that part of your body. Even closing your eyes and counting down from 10 is an effective method for reminding your body that this is not a fight or flight situation.

Another great way to prepare is to spend time imagining yourself succeeding. Called Creative Visualization (Gawain, 2016), some athletes use this technique before a game to improve their performance (Newmark, 2012). The following exercises can help you mentally prepare for your talk on the day you will be presenting:

Exercise 9.1 Belly Breathing

1. Sit in a chair, preferably in a quiet place (although this is not required).
2. Check your posture and make sure you are sitting tall but relaxed. Do not let yourself collapse into the back of the chair. A proper posture will help open the channels of your breathing.
3. Release and relax the muscles of your behind, allowing them to melt into the chair, but keep your pelvis in its neutral position (neither tipped forward nor back).
4. Place your hand on your belly. Notice your breath.
5. Using your thoughts and a release of muscles, let the breath drop down into your pelvis as you breathe. You should begin to feel the hand on your belly moving up and down. You might also feel your back grazing the back of the chair as your torso inflates. These are all good signs.
6. Give yourself at least 20 nice, slow belly breaths before returning to normal breathing.

Exercise 9.2 Tighten and Release

1. You may do this sitting in a chair, standing, or lying down. A quiet place is preferable but again not required.
2. Check your posture and apply proper alignment.
3. Move through your major muscles groups one by one, first tensing that muscle group as hard as possible, holding that tension for a moment, and then releasing that muscle group completely. For example, you could begin with all of the muscles in your right leg, then your left leg, then your pelvis, then your back, etc.
4. It can be tricky to do this without tensing other collections of muscles. Do your best to isolate the group you are tensing and to release any group that does not belong.
5. When you release, imagine the muscles melting away in relief—giving up the tension all at once.

Exercise 9.3 Creative Visualization

1. A few days prior to your talk, find a quiet space and either sit or lie down in a comfortable position.
2. Begin to imagine the process of giving your talk. Play the entire movie from introduction to exit.
3. Play the movie from an outsider's point of view, as though you are an audience member watching yourself present.
4. Now play the movie again from an insider's point of view, relishing the confidence and finesse that you are naturally presenting to an admiring crowd.
5. Make sure that your movie is positive and successful. This is not the time for morbid worrying. In these movies, everything goes perfectly.
6. Consider doing this exercise daily for one week before you present.

Creative Visualization involves practicing a mental movie of your talk: *Imagine yourself stepping confidently to the front of the room. Everything is perfectly prepared. Your equipment functions beautifully (because you checked it prior to the event). You begin with a bang; the audience is mesmerized. Your voice is open, resonant, and everyone can hear you. You make eye contact with the entire room, including a smile when appropriate, adjusting your delivery tempo to punctuate the important moments in your talk. When you reach the final lines of your presentation, you hold for a moment, and then button the talk with a confident "Thank you!" triggering enthusiastic applause. You answer follow-up questions with lucidity, clearly demonstrating your expertise. Then you gather your things and leave the stage with grace, your head held high.* Mentally imagining your victory can prime you for success. If Creative Visualization does not work for you, we recommend just faking it. Pretend to be calm and confident before, during, and after your presentation. "It doesn't really matter how nervous you are—*as long as you appear calm*" (Kushner, 2010, p. 124).

It might be helpful to place your preparation process on a timeline. Here is an example that you can adjust to meet your needs.

Six week preparation timeline

Week 1	• Sketch out the primary ideas for the talk • Establish your core message • Determine what images will anchor your talk
Week 2	• Determine the shape of the story or dramatic arc • Fill in the details and establish how you will create tension and resolution • Find a terrific beginning
Week 3	• Determine how to make a strong ending • Find moments of humor or surprise
Week 4	• Say your talk out loud at least three times without timing it • Make adjustments to your story and the visual aids as appropriate
Week 5	• Practice at least three times to get a sense of the performance of the talk. • Record your practice sessions and give yourself notes • Go live with your research talk by practicing on your own or in front of colleagues • Take notes from them and make adjustments

Week 6	• Practice your talk at least three times in costume with all props
	• Identify where you will move and how
	• Include the equipment check and the entrance in these three rehearsals
	• Practice your talk in full five times
Presentation day	• Check your equipment ahead of time
	• Find a place to relax and mentally prepare
	• Present what you have practiced and allow yourself to shine
	• If you make a mistake or two, learn from them and let them go
	• Focus on your accomplishments. You worked hard and prepared well
	• Celebrate after your talk!

As you prepare your research talk, remember that stage fright is your friend. Allow those nerves to provide some extra energy for your performance. Believe that you can do this and you will give a stunning presentation. You have done your homework: You have fully prepared and rehearsed your talk, you have checked your equipment, you are dressed appropriately, and you know your subject well. You are the world's expert on this particular topic. You know more about the content of your research talk than anyone in the room. So put on a brave face with a light smile and pretend the audience does not scare you. Remember they are just people. People have voluntarily come to hear you speak. They want you to do well. If you have prepared properly, you will.

10

How to Handle Questions

Most research talks provide time for a question and answer period once the talk has finished. For some speakers, this can be the most terrifying part of the presentation. Even if you would prefer not to take questions, it is important to save time for a question and answer session at the end of your talk if the venue in which you are presenting expects it. Think of a question and answer session as another opportunity to connect with your audience rather than a situation in which you will have to parry an attack. Most viewers who ask questions will be genuinely curious about something you have presented. Questions are a sign that you have sparked interest. Answering these questions allows you to further enhance your credibility as an expert on your particular research topic.

A typical question and answer session is between ten and twenty percent of your entire speaking time. You may need to reduce the content of your talk in order to accommodate questions within the time allotted for your presentation. Timing your question and answer session properly creates an impression of control. Staying in control will contribute to a perception of confidence, something that will assist you should you encounter a challenging question.

If you are concerned about your performance during the question and answer session, it might help to prepare by creating a collection of test questions ahead of time. You will not necessarily know which questions will be asked once you present, but even the process of anticipating questions can help you mentally prepare for the questions you will receive. Another way to prepare yourself would be to include a question and answer session during the practice sessions you present to peers. You can provide test questions to the colleagues who are helping you prepare, or allow them to construct their own.

© Springer Nature Switzerland AG 2019
I. Cohen, M. Dreyer-Lude, *Finding Your Research Voice*,
https://doi.org/10.1007/978-3-030-31520-7_10

You could also give this pretend audience various roles to play as difficult audience members, helping you learn to negotiate challenging questions.

Exercise 10.1 Preparing for a Question and Answer Session

1. Write a list of several questions that an audience member might ask about your research. Include a range of possibilities. Choose easy and difficult questions. Include questions that interrogate a data set or graph in your presentation. Include questions that challenge your hypothesis or indicate someone has already answered it. Include a question that asks about next steps in your research.
2. Now answer each of those questions verbally, practicing until you feel you have found a smooth and concise answer to each.

Participating in a question and answer session is a form of performance. There are some protocols and recommended best practices when answering questions in a formal situation.

Standard Question and Answer Protocol

1. When taking a question, *give the speaker your full attention*. Do not let yourself become distracted. Look the speaker in the eye and listen carefully. The audience is watching how you handle this moment.
2. *Listen carefully* to the speaker to make sure you are hearing what is being asked.
3. *Thank the speaker* for asking the question either by nodding in acknowledgment or by using one of the following scripts: "Thanks for asking that question." "That's a great question." "That's an interesting question." "I'm so glad you asked that question." Any script will do as long as it acknowledges the speaker in a positive way.
4. *Repeat the question back* to the speaker to be sure you have heard it correctly. "What I heard you ask is…" "You seem to be interested in more information about…" This is both respectful and can save time, as it will prevent you from answering a question that was not asked.
5. *Answer the question* to the best of your ability. Direct the answer to the speaker who asked the question, but be sure to look around and include the entire audience as well (Dale Carnegie Training, 2011). If you find yourself getting into an extended discussion with just one person and are leaving the rest of the audience behind, cut the discussion short and suggest meeting after the talk to continue the conversation.
6. *Verify* that you answered the question asked: "Does that answer your question?" The Dale Carnegie Institute training (2011) refers to this step as the "bridge" as it assists with a smooth transition from one question to the next.

In our experience the place where the question and answer session typically goes wrong is in step number 4, which requires determining the actual question. The following improvisation exercise will help you really listen to what is said and learn to build a logical response from that information.

Exercise 10.2 So What You Are Saying Is...So I'll...
1. Stand in a circle, or across from a partner if there are only two of you.
2. One person begins with a statement. Anything will do. Perhaps he or she begins with: "I'm feeling a little bit hungry."
3. The next person repeats what that person said precisely, beginning with the name of the game. Like this: "So what you are saying is that you're feeling a little bit hungry. So I'll order lunch for us at the pizzeria."
4. The next person takes the most recent addition to this narrative and adds to it something of their own: "So what you are saying is you'll order lunch for us at the pizzeria, so I'll call and make sure they have a table." And the next person: "So what you are saying is you'll call and see if they have a table for us, so I'll go and put on my shoes."
5. Listen carefully and repeat what you hear. Try and make sure that your addition to the narrative is logical and not a random statement. Others in the group can help guide these efforts.

Exercise 10.3 Practicing a Research Talk Question and Answer Session
1. If possible, gather several colleagues for a practice presentation session and include a question and answer session at the end. If you encounter a question that you were not expecting, prepare an answer for next time. Chances are that if one person had a question about a particular part of your presentation others will, too. If you can modify a future version of your talk to address that question, do so.
2. Consider asking your colleagues to play the role of a difficult questioner. Possibilities include:

   - A *hostile questioner* who asks something that challenges your credibility. For example, "Wasn't all this done by professor so-and-so 10 years ago?"
   - A *long-winded questioner* who wants to make a speech more than ask a question. In this case, try out some ways for respectfully interrupting the speaker. For example, "I'm sorry, but I seem to have missed your question."
   - A *dominating questioner* who insists on asking several questions in a row. Here, you might call on another questioner who has their hand up before the next question is asked.

3. During your actual presentation, you may find other points in the question and answer session during which you get stuck. Try to come up with an exercise for your next practice session that tackles these new issues.

There may be times giving a research talk when you have fully prepared, you are following proper protocol when taking questions, and things still seem to go awry. Below we list some last-resort strategies for handling difficult situations during such a session.

Tips for Handling Difficult Situations

1. *Aggressive or critical questions*: Individuals who present aggressive behavior may be looking for a fight. This may or may not be personal. The easiest way to diffuse this impulse is to find a way to agree with the individual (Meyers & Nix, 2011). For example: Individual: "Isn't your research just a rehash of X's investigation of Y?" You: "Thanks for asking that question. I do agree that what I'm doing could be perceived as a reexamination of already existing data. What's significant and unique about my work here is…"
2. *When you don't know the answer:* "I'd like to think about that a bit more before I answer it. Could you and I talk afterwards?" (Then, make sure to have the answer for next time!)
3. *When you disagree with the questioner:* "I appreciate your position on the issue. Perhaps you and I could discuss it in more detail later on."
4. *Someone seems to know a lot about your topic:* "You seem to know a lot about this subject. Perhaps you and I can compare notes after my talk is over."
5. In any difficult situation *maintain your cool*. Stay in control, keep a pleasant smile on your face, and do not let them upset you. Remember that most of the audience (perhaps all but one) is on your side.

We have provided a collection of best practices for question and answer sessions in this chapter. There are also behaviors you should avoid in order to have a successful question and answer session. Here are some recommendations for what not to do.

Things to Keep in Mind During a Question and Answer Session

1. *Don't interrupt* the speaker asking the question unless this person is making a speech (Dale Carnegie Training, 2011). You may think you understand what is being asked and are ready to answer long before the speaker has fully articulated the question. It is bad manners to interrupt unnecessarily, and you could be wrong. Be sure to check and make sure that you have understood the question properly.
2. Make sure you *answer the question that was asked* rather than the question that you wish had been asked (Dale Carnegie Training, 2011). Checking before you answer and once again when you have finished answering will help make sure you have answered their question rather than your own.

3. If you do not know the answer to the question *just say you do not know.* Even experts encounter moments when an audience member has asked something outside of their area of expertise (Collins, 2011). Not having all of the answers is perfectly all right if you handle it with grace. Express interest in knowing the answer yourself, and offer to research the answer and get back to them. That creates a win for both parties. You can reframe the situation when you encounter a question you cannot answer. "That's a really interesting question. We might be able to answer it by doing such and such experiments in the next phase of our research. Thanks for the suggestion."

4. Sometimes questioners are just mean. *Try not to take it personally* (PowerSpeaking, 2016) or get defensive. If they succeed in upsetting you, they win. If you maintain your dignity and gracefully deflect the question, you win. The audience will not appreciate watching someone attack you. If you handle the situation well, the person asking the hostile question will lose face and you will gain respect.

Answering questions is a bit of an art. Like the improvisation exercises in this book, it requires thinking on your feet and it gets easier with practice. Eventually, this part of the talk may become the most enjoyable as the audience reflects back to you how well you presented your research and how engaged they are in your discoveries.

11

Game Day

It is the day for the main event. You have important research to share and you have carefully prepared your research talk. Make this moment count. Get enough sleep the night before to make sure you are physically prepared. Dress nicely. It is always better to be overdressed than underdressed. Wearing nice clothes will help you feel confident and will convey respect for the audience. Finally, build yourself up emotionally. Be your own coach. You have been practicing your talk, perfecting your slides, and honing your performance skills. You have got this.

Here are some additional things to consider as you get ready to present your research.

Before You Arrive at the Site
- Make at least one backup copy of your slide show. You may want to bring your slides in multiple formats (in the Cloud, on a thumb drive, on your laptop). Having multiple copies will help put your mind at ease, should you encounter obstacles on site.
- Pack an extra outfit to wear. If your best outfit becomes stained or marred in travel, having a backup will make a difference. If you will be flying to the presentation site and you have a tight connection, wear your presentation clothes on the airplane. If your luggage is lost or delayed, you will still be well-dressed for the event.
- Remember to pack your power cords, your thumb drive, your media adapters (dongles), and your laser pointer/clicker. Although the organizers of your talk may have these supplies on site, it is always best to bring your own as a backup.

© Springer Nature Switzerland AG 2019
I. Cohen, M. Dreyer-Lude, *Finding Your Research Voice*,
https://doi.org/10.1007/978-3-030-31520-7_11

On the Day of Your Presentation
- If possible, visit the room where you will present several hours before your presentation time. Get a sense of the size and architecture of the space. Examine the audio-visual support you will use and take a good look at the microphone. Even doing this from a distance as you watch another speaker present in your space will help prepare you mentally.
- Arrive at least 30 min early for your presentation. If possible, find a quiet space to focus and relax before you present.
- When you are introduced, stand confidently and smile. Head to the front with a bounce in your step. Pause for a moment before you begin and take the audience in.
- Now present what you have rehearsed, forget about small mistakes, and enjoy sharing what you know with your colleagues.
- If possible, have a colleague record your presentation so that you can watch it at a later date.

After Your Presentation
- Smile and exit with your head held high.
- Try to stick around right after your talk in case someone wants to find you and ask further questions.
- If someone congratulates you on your talk, thank them. Do not be self-effacing or falsely modest. Even if you feel like you could have done better, the audience may have thought your talk was terrific.

It is best not to reflect on the details of your performance right away. It is hard to be objective right after a presentation: small mistakes become larger than life; big victories seem life changing. You can be more objective later, when the emotion of the event has dissipated. If you asked a colleague to record your performance, watch it several days after your presentation and objectively assess what worked and what did not. Remember that developing any skill takes time and practice. There is no such thing as a perfect talk. Each time we give a presentation is an opportunity to try out slight improvements. Master speakers may present the same talk 50 times before it is perfected. So be patient with yourself and acknowledge the improvements you make with each iteration.

Once your talk is over, it is time to celebrate by having dinner with friends or watching a movie. The celebration is an equally important part of the ritual. If you are giving a departmental seminar, this celebration may include a dinner or reception. After a conference talk, you should take the lead in planning your own celebration and go and have some fun. You have worked hard to construct your talk, and this effort needs to be acknowledged.

12

Conclusion

Becoming an expert public speaker is possible for anyone who is willing to put in the time and effort. If you have followed the advice in this book, you will probably have worked harder than you ever imagined to turn your own presentation into a clear, engaging, and entertaining performance. As your presentations improve, you will discover that the effort is well worth the reward.

Remember that there is no one way to give a good talk. What we have provided in this book is a set of exercises that will help guide you in turning your presentation into an engaging talk. This is just a beginning. We hope that as you improve in your presentation skills, the exercises here will become even more useful. As time goes on you may modify these exercises or even develop new ones that help perfect your best practices.

Throughout this book, we have encouraged you to break out of your comfort zone. It is easy to be ordinary when giving a research talk. You will see ordinary talks everywhere you go. The exercises in this book have been about teaching you to be extraordinary. To be exceptional, you must engage your imagination, develop some creative tactics to grab your audience's attention, and then take the risk of including those ideas in your presentation. Taking a risk in the creation of your talk and the presentation of your performance means that you might fail. You might tell a bad joke, try an exciting reveal that fizzles, or present a story that does not engage. But by risking these moments, you will learn what works and what does not, and you will adjust your choices accordingly. Chances are your audience will appreciate the fact that you are trying to provide something unexpected and interesting. You may find that risk-taking offers its own emotional reward as it can be exhilarating to take a chance when giving a talk, particularly when the risk pays off.

© Springer Nature Switzerland AG 2019
I. Cohen, M. Dreyer-Lude, *Finding Your Research Voice*,
https://doi.org/10.1007/978-3-030-31520-7_12

Giving a great talk to a fully engaged audience is thrilling. Once you experience the power to make them laugh on cue or get them to feel the things you want them to feel, you will find yourself inspired to continue to improve your presentation skills. As these skills mature and your ability to express your ideas flourishes, you will get a small taste of what makes being an actor so intoxicating… but that is another book.

References

Anholt, R. R. (2010). *Dazzle'em with style: The art of oral scientific presentation* (2nd ed.). Amsterdam: Elsevier.

Arnold, K. J. (2010). *Boring to bravo: Proven presentation techniques to engage, involve and inspire your audience to action.* Austin, TX: Greenleaf Book Group Press.

Burgoon, J. K., & Saine, T. P. (1978). *The unspoken dialogue: An introduction to nonverbal communication.* Boston: Houghton Mifflin Harcourt.

Cohen, I. (2019). Flight of the fruit fly. *Physical Review Fluids, 4*(11), 110503. https://doi.org/10.1007/978-3-030-31520-7_3

Collective Evolution. (2016). *6 Bad postures that are ruining your health and how to correct them*, March 11, 2016. https://www.collective-evolution.com/2016/03/11/6-bad-postures-that-are-ruining-your-health-how-to-correct-them/

Collins, P. (2011). *Speak with power and confidence.* New York: Sterling.

Conner, L. (2013). *Audience engagement and the role of arts talk in the digital era.* New York: Palgrave MacMillan.

Dale Carnegie Training. (2011). *Stand and deliver: How to become a masterful communicator and public speaker.* New York: Touchstone.

Doumont, J. [CTL Stanford]. (2013, April 4). *Creating effective slides: Design, use, and construction in science* [video file]. Retrieved from https://www.youtube.com/watch?v=meBXuTIPJQk&feature=youtu.be

Dresler, M., Shirer, W. R., Konrad, B. N., Müller, N. C., Wagner, I. C., Fernández, G., et al. (2017). Mnemonic training reshapes brain networks to support superior memory. *Neuron, 93*(5), 1227–1235.

Duarte, N. S. (2008). *Slide:ology: The art and science of creating great presentations.* Sebastopol, CA: O'Reilly Media.

Ekman, P. (2007). *Emotions revealed: Recognizing faces and feelings to improve*

© Springer Nature Switzerland AG 2019
I. Cohen, M. Dreyer-Lude, *Finding Your Research Voice*,
https://doi.org/10.1007/978-3-030-31520-7

communication and emotional life. London: Macmillan.

Feldenkrais, M. (2013). *Thinking and doing*. Longmont, CO: Genesis II Publishing.

Fiske, S. T. (2018). Stereotype content: Warmth and competence endure. *Current Directions in Psychological Science, 27*(2), 67–73.

Frankel, F., & DePace, A. H. (2012). *Visual strategies: A practical guide to graphics for scientists and engineers*. New Haven, CT: Yale University Press.

Frieder, R. E., Van Iddekinge, C. H., & Raymark, P. H. (2016). How quickly do interviewers reach decisions? An examination of interviewers' decision-making time across applicants. *Journal of Occupational and Organizational Psychology, 89*(2), 223–248.

Gallo, C. (2014). *Talk like TED: The 9 public speaking secrets of the world's top minds*. London: Pan Macmillan.

Gawain, S. (2016). *Creative visualization: Use the power of your imagination to create what you want in your life*. Novato, CA: New World Library.

Goldin-Meadow, S. (2005). *Hearing gesture: How our hands help us think*. Cambridge, MA: Harvard University Press.

Hall, E. T. (1963). A system for the notation of proxemic behavior. *American Anthropologist, 65*(5), 1003–1026.

Halpern, B. L., & Lubar, K. (2004). *Leadership presence*. London: Penguin Books.

Heath, C., & Heath, D. (2007). *Made to stick: why some ideas survive and others die*. New York: Random House.

Howard, K., & Tivnan, E. (2003). *Act natural: How to speak to any audience*. New York: Random House.

Howey, B. (2005). *The actor's menu: A character preparation handbook*. St. Petersburg, FL: Compass Publishing.

Jakubowicz, R. (2018). *The yoga mind: 52 essential principles of yoga philosophy to deepen your practice*. Berkeley, CA: Rockridge Press.

Kahneman, D. (2011). *Thinking, fast and slow*. London: Macmillan.

Key, M. R. (1975). *Paralanguage and kinesics (nonverbal communication)*. Lanham, MD: Scarecrow Press.

Knaflic, C. N. (2015). *Storytelling with data: A data visualization guide for business professionals*. Hoboken, NJ: Wiley.

Kushner, M. (2010). *Public speaking for dummies*. Hoboken, NJ: Wiley.

Leitman, M. (2015). *Long story short*. Seattle, WA: Sasquatch Books.

Linklater, K. (1976). *Freeing the natural voice*. New York: Drama Book Publishers.

MacPhail, T. (2015). *The personal touch: Using anecdotes to hook a reader. Conciliate*. https://chroniclevitae.com/news/1019-the-personal-touch-using-anecdotes-to-hook-a-reader

Maisel, E. (1989). *The Alexander technique: The essential writings of F. Matthias Alexander*. St. Louis, MO: Carol Publishing Group.

McKee, R. (1997). *Story: Style, structure, substance, and the principles of screenwriting*. New York: Harper Collins.

Menzel, K. E., & Carrell, L. J. (1994). The relationship between preparation and performance in public speaking. *Communication Education, 43*(1), 17–26.

Meyers, P., & Nix, S. (2011). *As we speak: How to make your point and have it stick.* New York: Simon and Schuster.

Monarth, H. (2009). *Executive presence: The art of commanding respect like a CEO.* New York: McGraw Hill Professional.

Newmark, T. (2012). Cases in visualization for improved athletic performance. *Psychiatric Annals, 42*(10), 385–387.

Niedenthal, P. M. (2007). Embodying emotion. *Science, 316*(5827), 1002–1005.

Oster, E. (2007). *Flip your thinking on AIDS in Africa* [video file]. Retrieved from https://www.ted.com/talks/emily_oster_flips_our_thinking_on_aids_in_africa?referrer=playlist-everything_you_thought_was

Phillips, D. J. P. [TEDx Talks]. (2014, April 14). *How to avoid death by powerpoint* [video file]. Retrieved from https://www.youtube.com/watch?v=Iwpi1Lm6dFo

PowerSpeaking Inc. (2016, Nov 30). *Tough questions: Tips for dealing with difficult audiences.* Retrieved from https://blog.powerspeaking.com/tough_questions_tips_for_dealing_with_difficult_audiences

Reynolds, G. (2012). *Presentation Zen: Simple ideas on presentation design and delivery.* San Francisco: New Riders.

Rodenburg, P. (2015). *The actor speaks: Voice and the performer.* New York: St. Martin's Griffin.

Sarnoff, D., & Moore, G. (1987). *Never be nervous again.* New York: Crown Publishers.

Scheflen, A. E. (1972). *Body language and the social order: Communication as behavioral control.* Upper Saddle River, NJ: Prentice Hall.

Stone, J., & Bachner, J. (1994). *Speaking up: A book for every woman who wants to speak effectively.* New York: McGraw-Hill Companies.

Thomas, J. (2013). *Script analysis for actors, directors, and designers.* Boca Raton, FL: CRC Press.

Weissman, J. (2008). *Presenting to win: The art of telling your story.* Upper Saddle River, NJ: FT Press.

Zak, P. (2012). *How the dramatic arc can change our brain chemistry and spur us to action.* https://futureofstorytelling.org/video/paul-zak-empathy-neurochemistry-and-the-dramatic-arc

Zeoli, R. (2008). *The 7 principles of public speaking: Proven methods from a PR professional.* New York: Skyhorse Publishing.

Printed in the United States
By Bookmasters